HUMAN UNIVERSE

Professor Brian Cox
& Andrew Cohen

WILLIAM
COLLINS

WHAT A PIECE OF WORK IS A
MAN, HOW NOBLE IN REASON,
HOW INFINITE IN FACULTIES,
IN FORM AND MOVING HOW
EXPRESS AND ADMIRABLE, IN
ACTION HOW LIKE AN ANGEL, IN
APPREHENSION HOW LIKE A GOD!
THE BEAUTY OF THE WORLD,
THE PARAGON OF ANIMALS —
AND YET, TO ME, WHAT IS THIS
QUINTESSENCE OF DUST? MAN
DELIGHTS NOT ME — NOR WOMAN
NEITHER, THOUGH BY YOUR
SMILING YOU SEEM TO SAY SO.

HAMLET

What is a human being? Objectively, nothing of consequence. Particles of dust in an infinite arena, present for an instant in eternity. Clumps of atoms in a universe with more galaxies than people. And yet a human being is necessary for the question itself to exist, and the presence of a question in the universe – any question – is the most wonderful thing. Questions require minds, and minds bring meaning. What is meaning? I don't know, except that the universe and every pointless speck inside it means something to me. I am astonished by the existence of a single atom, and find my civilisation to be an outrageous imprint on reality. I don't understand it. Nobody does, but it makes me smile.

This book asks questions about our origins, our destiny, and our place in the universe. We have no right to expect answers; we have no right to even ask. But ask and wonder we do. *Human Universe* is first and foremost a love letter to humanity; a celebration of our outrageous fortune in existing at all. I have chosen to write my letter in the language of science, because there is no better demonstration of our magnificent ascent from dust to paragon of animals than the exponentiation of knowledge generated by science. Two million years ago we were apemen. Now we are spacemen. That has happened, as far as we know, nowhere else. That is worth celebrating.

WHERE ARE WE?

We shall not cease from exploration,
And the end of all our exploring
Will be to arrive where we started
And know the place for the first time.

T.S. Eliot

OAKBANK AVENUE, CHADDERTON, OLDHAM, GREATER MANCHESTER, ENGLAND, UNITED KINGDOM, EUROPE, EARTH, MILKY WAY, OBSERVABLE UNIVERSE ... ?

FROM THE EARTH TO THE SUN

1 astronomical unit (AU)

FROM EARTH TO NEAREST STAR
265,000 AU

FROM EARTH TO CENTRE OF MILKY WAY
1,580,000,000 AU

FROM EARTH TO ANDROMEDA
1,580,000,000,000 AU

FROM EARTH TO FARTHEST GALAXY
8,532,000,000,000,000 AU

For me, it was an early 1960s brick-built bungalow on Oakbank Avenue. If the wind was blowing from the east you could smell vinegar coming from Sarson's Brewery – although these were rare days in Oldham, a town usually subjected to Westerlies dumping Atlantic moisture onto the textile mills, dampening their red brick in a permanent sheen against the sodden sky. On a good day, though, you'd take the vinegar in return for sunlight on the moors. Oldham looks like Joy Division sounds – and I like Joy Division. There was a newsagent on the corner of Kenilworth Avenue and Middleton Road and on Fridays my granddad would take me there and we'd buy a toy – usually a little car or truck. I've still got most of them. When I was older, I'd play tennis on the red cinder courts in Chadderton Hall Park and drink Woodpecker cider on the bench in the grounds of St Matthew's Church. One autumn evening just after the start of the school year, and after a few sips, I had my first kiss there – all cold nose and sniffles. I suppose that sort of behaviour is frowned upon these days; the bloke in the off-licence would have been prosecuted by Oldham Council's underage cider tsar and I'd be on a list. But I survived, and, eventually, I left Oldham for the University of Manchester.

Everyone has an Oakbank Avenue; a place in space at the beginning of our time, central to an expanding personal universe. For our distant ancestors in the East African Rift, their expansion was one of physical experience alone, but for a human fortunate to be born in the latter half of the twentieth century in a country like mine, education powers the mind beyond direct experience – onwards and outwards and, in the case of this little boy, towards the stars.

As England stomped its way through the 1970s, I learned my place amongst the continents and oceans of our blue planet. I could tell you about polar bears on Arctic ice flows or gazelle grazing on central plains long before I physically left our shores. I discovered that our Earth is one planet amongst nine (now redefined as eight) tracing out an elliptical orbit around an average star, with Mercury and Venus on the inside and Mars, Jupiter, Saturn, Uranus and Neptune beyond. The Sun is one star amongst 400 billion in the Milky Way Galaxy, itself just one galaxy amongst 350 billion in the observable universe. Later, at university, I discovered that physical reality extends way beyond the 90-billion-light-year visible sphere into – if I had to guess based on my 46-year immersion in the combined knowledge of human civilisation – infinity.

This is my ascent into insignificance; a road travelled by many and yet one that remains intensely personal to each individual who takes it. The routes we follow through the ever-growing landscape of human knowledge are chaotic; the delayed turn of a page in a stumbled-upon book can lead to a lifetime of exploration. But there are common themes amongst our disparate intellectual journeys, and the relentless relegation from centre stage that inevitably followed the development of modern astronomy has had a powerful effect on our shared experience. I am certain that the voyage from the centre of creation to an infinitesimally tiny speck should be termed an ascent, the most glorious intellectual climb. Of course, I also recognise that there are many who have struggled – and continue to struggle – with such a dizzying physical relegation.

John Updike once wrote that 'Astronomy is what we have now instead of theology. The terrors are less, but the comforts are nil'. For me, the choice between fear and elation is a matter of perspective, and it is a central aim of this book to make the case for elation. This may appear

THE ROLE OF THE CREATOR
In Rome I visited the Vatican Observatory (established in 1787), one of the oldest astronomical research institutions in the world.

at first sight to be a difficult challenge – the very title *Human Universe* appears to demonstrate an unjustifiable solipsism. How can a possibly infinite reality be viewed through the prism of a bunch of biological machines temporarily inhabiting a mote of dust? My answer to that is that *Human Universe* is a love letter to humanity, because our mote of dust is the only place where love certainly exists.

This sounds like a return to the anthropocentric vision we held for so long, and which science has done so much to destroy in a million humble cuts. Perhaps. But let me offer an alternative view. There is only one corner of the universe where we know for sure that the laws of nature have conspired to produce a species capable of transcending the physical bounds of a single life and developing a library of knowledge beyond the capacity of a million individual brains which contains a precise description of our location in space and time. We know our place, and that makes us valuable and, at least in our local cosmic neighbourhood, unique. We don't know how far we would have to travel to find another such island of understanding, but it is surely a long long way. This makes the human race worth celebrating, our library worth nurturing, and our existence worth protecting.

Building on these ideas, my view is that we humans represent an isolated island of meaning in a meaningless universe, and I should immediately clarify what I mean by meaningless. I see no reason for the existence of the universe in a teleological sense; there is surely no final cause or purpose. Rather, I think that meaning is an emergent property; it appeared on Earth when the brains of our ancestors became large enough to allow for primitive culture – probably between 3 and 4 million years ago with the emergence of Australopithecus in the Rift Valley. There are surely other intelligent beings in the billions of galaxies beyond the Milky Way, and if the modern theory of eternal inflation is correct, then there is an infinite number of inhabited worlds in the multiverse beyond the horizon. I am much less certain that there are large numbers of civilisations sharing our galaxy, however, which is why I use the term 'isolated'. If we are currently alone in the Milky Way, then the vast distances between the galaxies probably mean that we will never get to discuss our situation with anyone else.

We will encounter all these ideas and arguments later in this book, and I will carefully separate my opinion from that of science – or rather what we know with a level of certainty. But it is worth noting that the modern picture of a vast and possibly infinite cosmos, populated with uncountable worlds, has a long and violent history, and the often-visceral reaction to the physical demotion of humanity lays bare deeply held prejudices and comfortable assumptions that sit, perhaps, at the core of our being. It seems appropriate, therefore, to begin this tour of the human universe with a controversial figure whose life and death resonates with many of these intellectual and emotional challenges.

Giordano Bruno is as famous for his death as for his life and work. On 17 February 1600, his tongue pinioned to prevent him from repeating his heresy (which recalls the stoning scene in Monty Python's *Life of Brian* when the admonishment 'you're only making it worse for yourself' is correctly observed to be an empty threat), Bruno was burned at the stake in the Campo de' Fiori in Rome and his ashes thrown into the Tiber. His crimes were numerous and included heretical ideas such as denying the divinity of Jesus. It is also the opinion of many historians that Bruno was irritating, argumentative and, not to put too fine a point on it, an all-round pain in the arse, so many powerful people were simply glad to see the back of him. But it is also true that Bruno embraced and promoted a

wonderful idea that raises important and challenging questions. Bruno believed that the universe is infinite and filled with an infinite number of habitable worlds. He also believed that although each world exists for a brief moment when compared to the life of the universe, space itself is neither created nor destroyed; the universe is eternal.

Although the precise reasons for Bruno's death sentence are still debated amongst historians, the idea of an infinite and eternal universe seems to have been central to his fate, because it clearly raises questions about the role of a creator. Bruno knew this, of course, which is why his return to Italy in 1591 after a safe, successful existence in the more tolerant atmosphere of northern Europe remains a mystery. During the 1580s Bruno enjoyed the patronage of both King Henry III of France and Queen Elizabeth I of England, loudly promoting the Copernican ideal of a Sun-centred solar system. Whilst it's often assumed that the very idea of removing the Earth from the centre of the solar system was enough to elicit a violent response from the Church, Copernicanism itself was not considered heretical in 1600, and the infamous tussles with Galileo lay thirty years in the future. Rather, it was Bruno's philosophical idea of an eternal universe, requiring no point of creation, which unsettled the Church authorities, and perhaps paved the way for their later battles with astronomy and science. As we shall see, the idea of a universe that existed before the Big Bang is now central to modern cosmology and falls very much within the realm of observational and theoretical science. In my view this presents as great a challenge to modern-day theologians as it did in Bruno's time, so it's perhaps no wonder that he was dispensed with.

Bruno, then, was a complex figure, and his contributions to science are questionable. He was more belligerent free-thinker than proto-scientist, and whilst there is no shame in that, the intellectual origins of our ascent into insignificance lie elsewhere. Bruno was a brash, if portentous, messenger who would likely not have reached his heretical conclusions about an infinite and eternal universe without the work of Nicolaus Copernicus, grounded in what can now clearly be recognised as one of the earliest examples of modern science, and published over half a century before Bruno's cinematic demise.

BRUNO'S HERETICAL SCIENCE
This bas-relief shows Giordano Bruno (1548–1600) being burned at the stake for his heretical and revolutionary ideas, among which was his belief that the universe is infinite and contained numerous habitable worlds.

VATICAN OBSERVATORY
The Vatican Observatory is based in Castel Gandolfo, the pope's summer residence outside Rome.

OFF CENTRE

Nicolaus Copernicus was born in the Polish city of Torun in 1473 and benefited from a superb education after being enrolled at the University of Cracow at 18 by his influential uncle, the Bishop of Warmia. In 1496, intending to follow in the footsteps of his uncle, Copernicus moved to Bologna to study canon law, where he lodged with an astronomy professor, Domenica Maria de Novara, who had a reputation for questioning the classical works of the ancient Greeks and in particular their widely accepted cosmology.

The classical view of the universe was based on Aristotle's not unreasonable assertion that the Earth is at the centre of all things, and that everything moves around it. This feels right because we don't perceive ourselves to be in motion and the Sun, Moon, planets and stars appear to sweep across the sky around us. However, a little careful observation reveals that the situation is in fact more complicated than this. In particular, the planets perform little loops in the sky at certain times of year, reversing their track across the background stars before continuing along their paths through the constellations of the zodiac. This observational fact, which is known as retrograde motion, occurs because we are viewing the planets from a moving vantage point – the Earth – in orbit around the Sun.

This is by far the simplest explanation for the evidence, although it is possible to construct a system capable of predicting the position of the planets months or years ahead and maintain Earth's unique stationary position at the centre of all things. Such an Earth-centred model was developed by Ptolemy in the second century and published in his most famous work, *Almagest*. The details are extremely complicated, and aren't worth describing in detail here because the central idea is totally wrong and we won't learn anything. The sheer contrived complexity of an Earth-centred description of planetary motions can be seen in the illustration opposite, which shows the apparent motions of the planets against the stars as viewed from Earth. This tangled Ptolemaic system of Earth-centred circular motions, replete with the arcane terminology of epicycles, deferents and equant, was used successfully by astrologers for thousands of years to predict where the planets would be against the constellations of the zodiac – presumably allowing them to write their horoscopes and mislead the gullible citizens of the ancient world. And if all you care about are the predictions themselves, and your philosophical prejudice and common-sense feeling of stillness require the Earth to be at the centre, then everything is fine. And so it remained until Copernicus became sufficiently offended by the sheer ugliness of the Ptolemaic model to do something about it.

Copernicus's precise objections to Ptolemy are not known, but sometime around 1510 he wrote an unpublished manuscript called the *Commentariolus* in which he expressed his dissatisfaction with the model. 'I often considered whether there could perhaps be found a more reasonable arrangement of circles, from which every apparent irregularity would be derived while everything in itself would move uniformly, as is required by the rule of perfect motion.'

The *Commentariolus* contained a list of radical and mostly correct assertions. Copernicus wrote that the Moon revolves around the Earth, the planets rotate around the Sun, and the distance from the Earth to the Sun is an insignificant fraction of the distance to the stars. He was the first to suggest that the Earth rotates on its axis, and that this rotation is responsible for the daily motion of the Sun and stars across the sky. He also understood that the retrograde motion of the planets is due to the motion of the Earth and not the planets themselves. Copernicus

always intended *Commentariolus* to be the introduction to a much larger work, and included little if any detail about how he had come upon such a radical departure from classical ideas. The full justification for and description of his new cosmology took him a further 20 years, but by 1539 he had finished most of his six-volume *De revolutionibus*, although the completed books were not published until 1543. They contained the mathematical elaborations of his heliocentric model, an analysis of the precession of the equinoxes, the orbit of the Moon, and a catalogue of the stars, and are rightly regarded as foundational works in the development of modern science. They were widely read in universities across Europe and admired for the accuracy of the astronomical predictions contained within. It is interesting to note, however, that the intellectual turmoil caused by our relegation from the centre of all things still coloured the view of many of the great scientific names of the age. Tycho Brahe, the greatest astronomical observer before the invention of the telescope, referred to Copernicus as a second Ptolemy (which was meant as a compliment), but didn't accept the Sun-centred solar system model in

PTOLEMY'S MODEL
This diagram shows the apparent motions of the Sun and planets across the sky as seen from the perspective of the Earth.

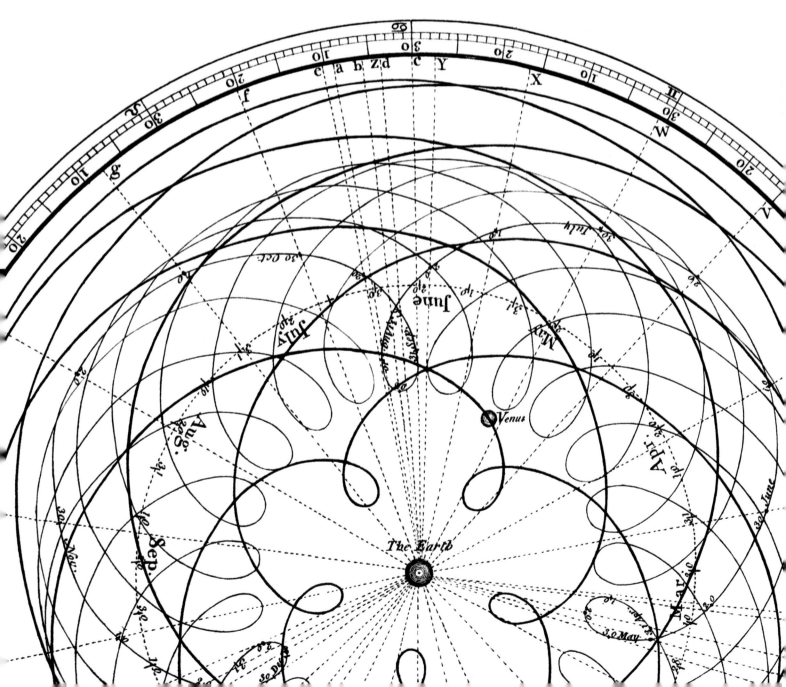

COPERNICUS'S HELIOCENTRIC MODEL
In 1510 Copernicus rejected the Ptolomeic model in his work the *Commentariolus*, and instead suggested that the Moon revolved around the Earth and the planets around the Sun. Many other assertions in this work proved to be largely correct, including his radical idea that the Earth rotated on an axis and was responsible for the daily motion of the Sun and stars across the sky.

its entirety, partly because he perceived it to be in contradiction with the Bible, but partly because it does seem obvious that the Earth is at rest. This is not a trivial objection to a Copernican solar system, and a truly modern understanding of precisely what 'at rest' and 'moving' mean requires Einstein's theories of relativity – which we will get to later! Even Copernicus himself was clear that the Sun still rested at the centre of the universe. But as the seventeenth century wore on, precision observations greatly improved due to the invention of the telescope and an increasingly mature application of mathematics to describe the data, and led a host of astronomers and mathematicians – including Johannes Kepler, Galileo and ultimately Isaac Newton – towards an understanding of the workings of the solar system. This theory is good enough even today to send space probes to the outer planets with absolute precision.

At first sight it is difficult to understand why Ptolemy's contrived mess lasted so long, but there is a modern bias to this statement that is revealing. Today, a scientifically literate person assumes that there is a real, predictable universe beyond Earth that operates according to laws of

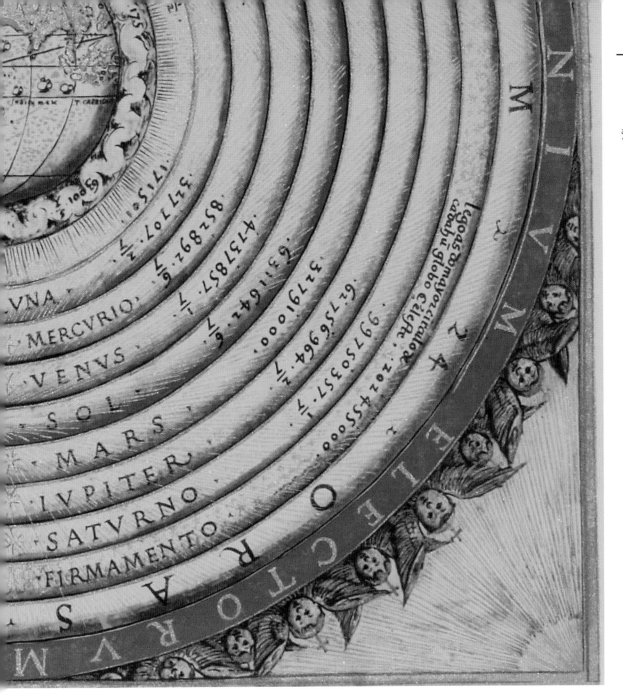

nature – the same laws that objects obey here on Earth. This idea, which is correct, only emerged fully formed with the work of Isaac Newton in the 1680s, over a century after Copernicus. Ancient astronomers were interested primarily in predictions, and although the nature of physical reality was debated, the central scientific idea of universal laws of physics had simply not been discovered. Ptolemy created a model that makes predictions that agree with observation to a reasonable level of accuracy, and that was good enough for most people. There had been notable dissenting voices, of course – the history of ideas is never linear. Epicurus, writing around 300 BCE, proposed an eternal cosmos populated by an infinity of worlds, and around the same time Aristarchus proposed a Sun-centred universe about which the Earth and planets orbit. There was also a strong tradition of classic orthodoxy in the Islamic world in the tenth and eleventh centuries. The astronomer and mathematician Ibn al-Haytham pointed out that, whilst Ptolemy's model had predictive power, the motions of the planets as shown in the figure on page 19 represent 'an arrangement that is impossible to exist'.

NEWTON'S LAW OF GRAVITY

F
Force between the masses

G
Gravitational constant

m₁
First mass

m₂
Second mass

r
Distance between the centres of the masses

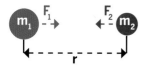

$$F_1 = F_2 = G\,\frac{m_1 \times m_2}{r^2}$$

The end of the revolution started by Copernicus around 1510, and the beginning of modern mathematical physics can be dated to 5 July 1687, when Isaac Newton published the *Principia*. He demonstrated that the Earth-centred Ptolemaic jumble can be replaced by a Sun-centred solar system and a law of universal gravitation, which applies to all objects in the universe and can be expressed in a single mathematical equation:

$$F = G\,\frac{m_1 m_2}{r^2}$$

The equation says that the gravitational force between two objects – a planet and a star, say – of masses m_1 and m_2 can be calculated by multiplying the masses together, dividing by the square of the distance r between them, and multiplying by G, which encodes the strength of the gravitational force itself. G, which is known as Newton's Constant, is, as far as we know, a fundamental property of our universe – it is a single number which is the same everywhere and has remained so for all time. Henry Cavendish first measured G in a famous experiment in 1798, in which he managed (indirectly) to measure the gravitational force between lead balls of known mass using a torsion balance. This is yet another example of the central idea of modern physics – lead balls obey the same laws of nature as stars and planets. For the record, the current best measurement of $G = 6.67 \times 10^{-11}$ N m²/kg², which tells you that the gravitational force between two balls of mass 1kg each, 1 metre apart, is just less than ten thousand millionths of a Newton. Gravity is a very weak force indeed, and this is why its strength was not measured until 71 years after Newton's death.

This is a quite brilliant simplification, and perhaps more importantly, the pivotal discovery of the deep relationship between mathematics and nature which underpins the success of science, described so eloquently by the philosopher and mathematician Bertrand Russell: 'Mathematics, rightly viewed, possesses not only truth, but supreme beauty – a beauty cold and austere, like that of sculpture, without appeal to any part of our weaker nature, without the gorgeous trappings of painting or music, yet sublimely pure, and capable of a stern perfection such as only the greatest art can show. The true spirit of delight, the exaltation, the sense of being more than Man, which is the touchstone of the highest excellence, is to be found in mathematics as surely as in poetry.'

Nowhere is this sentiment made more clearly manifest than in Newton's Law of Gravitation. Given the position and velocity of the planets at a single moment, the geometry of the solar system at any time millions of years into the future can be calculated. Compare that economy – you could write all the necessary information on the back of an envelope – with Ptolemy's whirling offset epicycles. Physicists greatly prize such economy; if a large array of complex phenomena can be described by a simple law or equation, this usually implies that we are on the right track.

The quest for elegance and economy in the description of nature guides theoretical physicists to this day, and will form a central part of our story as we trace the development of modern cosmology. Seen in this light, Copernicus assumes even greater historical importance. Not only did he catalyse the destruction of the Earth-centred cosmos, but he inspired Brahe, Kepler, Galileo, Newton and many others towards the development of modern mathematical physics – which is not only remarkably successful in its description of the universe, but was also necessary for the emergence of our modern technological civilisation. Take note, politicians, economists and science policy advisors of the twenty-first century; a prerequisite for the creation of the intellectual edifice upon which your spreadsheets, air-conditioned offices and mobile phones rest was the curiosity-driven quest to understand the motions of the planets and the Earth's place amongst the stars.

AT THE CENTRE OF THE SOLAR SYSTEM

Matching the observations of the wandering stars – the planets – of the night sky with the idea that the Earth was at the centre of the solar system required extremely complex models. In the case of Venus, combining the Earth at the centre with the observations meant that Venus had a circular orbit around a point midway between the Earth and the Sun, so-called epicycles, with all the other planets having similar complicated orbits around various points scattered around the solar system. Placing the Sun at the centre of the solar system, with the planets arranged in their familiar order, with the Moon orbiting the Earth, gave a much simpler system.

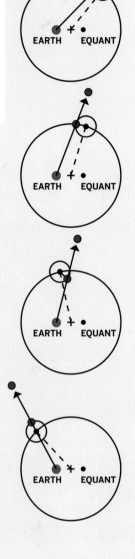

PTOLEMY'S SYSTEM

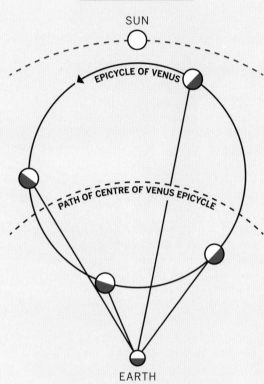

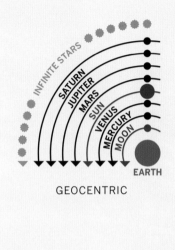

GEOCENTRIC

COPERNICUS'S SYSTEM

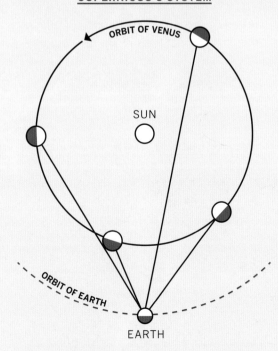

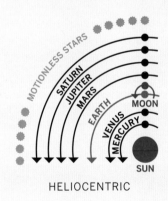

HELIOCENTRIC

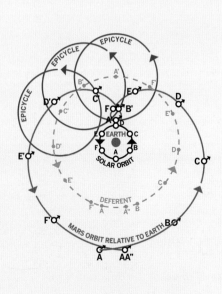

CHANGING PERSPECTIVE

Borman: Oh my God! Look at that picture over there.
Here's the Earth coming up. Wow, is that pretty.
Anders: Hey, don't take that, it's not scheduled.
Borman: (laughing) You got a color film, Jim?
Anders: Hand me that roll of color quick, will you...?
Lovell: Oh, man, that's great!

1968 was a difficult year on planet Earth. The Vietnam War, the bloodiest of Cold War proxy tussles, was at its height, ultimately claiming over three million lives. Martin Luther King Jr. was assassinated in Memphis, prompting presidential hopeful Bobby Kennedy to ask the people of the United States 'to tame the savageness of man and make gentle the life of this world.' Kennedy himself was assassinated before the year was out. Elsewhere, Russian tanks rolled into Prague, and France teetered on the edge of revolution. As I approached my first Christmas, my parents could have been forgiven for wondering what kind of world their son would inhabit in 1969. And then, as Christmas Eve drifted into Christmas morning, an unexpected snowfall decorated Oakbank Avenue and Borman, Lovell and Anders, 400,000 kilometres away, saved 1968.

Apollo 8 was, in the eyes of many, the Moon mission that had the most profound historical impact. It was a terrific, noble risk; a magnificent roll of the dice; a distillation of all that is great about exploration; a tribute to the sheer balls of the astronauts and engineers who decided that, come what may, they would honour President Kennedy's pledge to send 'a giant rocket more than 300 feet tall, the length of this football field, made of new metal alloys, some of which have not yet been invented, capable of standing heat and stresses several times more than have ever been experienced, fitted together with a precision better than the finest watch, carrying all the equipment needed for propulsion, guidance, control, communications, food and survival, on an untried mission, to an unknown celestial body, and then return it safely to Earth, re-entering the atmosphere at speeds of over 40,000 kilometres per hour, causing heat about half that of the temperature of the Sun – almost as hot as it is here today – and do all this, and do it right, and do it first before this decade is out'. If I heard that from a leader today I'd be first on the rocket. Instead I have to listen to vacuous diatribes about 'fairness', 'hard-working families', and how 'we're all in it together'. Sod that, I want to go to Mars.

To set Apollo 8 in context, Apollo 7, the first manned test flight of the Apollo programme, was flown by Schirra, Eisele and Cunningham in October 1968. Apollo 8 was supposed to be a December test flight for the Lunar Lander, conducted in the familiar surroundings of Earth orbit, but delivery delays meant that it was not ready for flight and the aim of meeting Kennedy's deadline looked to be dead. But this wasn't the twenty-first century, it was the 1960s and NASA was run by engineers. The programme manager was George Low, an army veteran and aeronautical engineer who knew the spacecraft inside out and had the strength of character to make decisions. Why not send Apollo 8 directly to the Moon without the Lunar Lander, proposed Low, allowing Apollo 9 to test-fly the LEM (Lunar Excursion Module) in Earth orbit in early 1969 when it became available and pave the way for a landing before the decade was out? Virtually every engineer at NASA is said to have agreed, and so it was that only the second manned flight of the Apollo spacecraft lifted off from Kennedy on 21 December, ten short weeks after Apollo 7, bound for the Moon. The crew later said that they estimated their chance of succeeding to be fifty-fifty.

Precisely 69 hours, 8 minutes and 16 seconds after launch, the Command Module's engine fired to slow the spacecraft down and allow

EARTHRISE, APOLLO 8
This famous image was taken by US astronauts on board the Apollo 8 spacecraft on 24 December 1968 as they orbited the Moon. This photograph has become iconic for its depiction of the beauty and fragility of the Earth.

it to be captured by the Moon's gravity, putting the three astronauts into lunar orbit. Newton's almost three-hundred-year-old equations were used to calculate the trajectory. This was a spectacular, practically unbelievable engineering triumph. Less than a decade after Yuri Gagarin became the first human to orbit the Earth, three astronauts travelled all the way to the Moon. But the mission's powerful and enduring cultural legacy rests largely on two very human actions by the crew. One was the famous and moving Christmas broadcast, the most-watched television event in history at that time, when distant explorers read the first lines from the Book of Genesis; 'We are now approaching lunar sunrise, and for all the people back on Earth, the crew of Apollo 8 has a message that we would like to send to you,' began Anders. 'In the beginning God created the heaven and the Earth. And the Earth was without form, and void; and darkness was upon the face of the deep.' Borman concluded with a sentence clearly spoken by a lonely man 400,000 kilometres from home. 'And from the crew of Apollo 8, we close with goodnight, good luck, a Merry Christmas – and God bless all of you, all of you on the good Earth.'

The mission's most potent legacy, however, is NASA image AS8-14-2383, snapped by Bill Anders on a Hasselblad 500 EL at f/11 and a shutter speed of 1/250th of a second on Kodak Ektachrome film. It was, in other words, a very bright photograph. The image is better known as Earthrise. When viewed with the lunar surface at the bottom, Earth is tilted on its side with the South Pole to the left, and the Equator running top to bottom. Little landmass can be seen through the swirling clouds, but the bright sands of the Namib and Saharan deserts stand out salmon pink against the blackness beyond. Just 368 years and ten months after a man was burned at the stake for dreaming of worlds without end, here is Earth, a fragile crescent suspended over an alien landscape, the negative of a waxing Moon in the friendly skies of Earth. This is an unfamiliar, planetary Earth, no longer central; just another world. When Kennedy spoke of Apollo as a journey to an unknown celestial body, he meant the Moon. But we discovered Earth and, in the words of T.S. Eliot, came to know the place for the first time.

OUTWARDS TO THE MILKY WAY

Newton's laws are the keys to understanding our place in our local neighbourhood. Coupled with precision observations of the motion of the planets and moons, they allow the scale and geometry of the solar system to be deduced, and their positions to be calculated at any point in the future. The nature and location of the stars, however, requires an entirely different approach because at first sight they appear to be point-like and fixed. The observation that the stars don't appear to move is important if you know something about parallax, as the ancients did. Parallax is the name given to a familiar effect. Hold your finger up in front of your face and alternately close each of your eyes, keeping your finger still. Your finger appears to move relative to the more distant background, and the closer your finger is to your face, the more it appears to move. This is not an optical illusion; it's a consequence of viewing a nearby object from two different spatial positions; in this case the two slightly different positions of your eyes. We don't normally perceive this parallax effect because the brain combines the inputs from the eyes to create a single image, although the information is exploited to create our sense of depth. Aristotle used the lack of stellar parallax to argue that the Earth must be stationary at the centre of the universe, because if the Earth moved then the nearby stars would be observed to move against the background of the more distant ones. Thousands of years later,

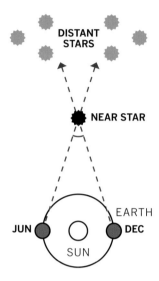

STELLAR PARALLAX

Nearby stars appear to move with respect to more distant stars due to the motion of the Earth around the Sun.

The line of sight to the star in December is different than that in June, when the Earth is on the other side of its orbit. Seen from Earth, the nearby star appears to sweep through the angle shown.

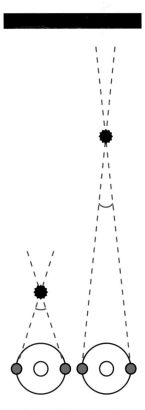

As the distance to a star increases, its parallax decreases. In the left figure, the star is about 2.5 times nearer than the star in the right figure, and has a parallax angle 2.5 times larger.

Tycho Brahe used a similar argument to refute the conclusions reached by Copernicus. Their logic was completely sound, but the conclusion is wrong because the nearby stars *do* move relative to the more distant background stars as the Earth orbits the Sun, and indeed as the Sun orbits the galaxy itself. You just have to look extremely carefully to see the effect.

Amongst the thousands of stars visible to the naked eye, 61 Cygni is one of the faintest. It's not without interest, being a binary star system of two orange K-type dwarf stars, slightly smaller and cooler than the Sun, orbiting each other at the lethargic rate of around 700 years. Despite the pair's relative visual anonymity, however, 61 Cygni has great historical significance. The reason for this quiet fame is that this faint star system was the first to have its distance from Earth measured by parallax.

Friedrich Bessel is best known to a physicist or mathematician for his work on the mathematical functions that bear his name. Pretty much any engineering or physical problem that involves a cylindrical or spherical geometry ends up with the use of Bessel functions, and, in blissful ignorance, you will probably encounter some piece of technology that has relied on them in the design process at some point today. But Bessel was first and foremost an astronomer, being appointed director of the Königsberg Observatory at the age of only 25. In 1838, Bessel observed that 61 Cygni shifted its position in the sky by approximately two-thirds of an arcsecond over a period of a year as viewed from Earth. That's not very much – an arcsecond is one 3600th of a degree. It is enough, however, to do a bit of trigonometry and calculate that 61 Cygni is 10.3 light years away from our solar system. This compares very favourably with the modern measurement of the distance, 11.41 ± 0.02 light years. Parallax is so important in astronomy that there is a measurement system completely based on it, which allows you to do these sums in your head. Astronomers use a distance measurement known as a parsec – which stands for 'per arcsecond'. This is the distance of a star from the Sun that has a parallax of 1 arcsecond. 1 parsec is 3.26 light years. Bessel's measurement of the parallax of 61 Cygni was 0.314 arcseconds, and this immediately implies that it's around 10 light years away.

Even today, stellar parallax remains the most accurate way of determining the distance to nearby stars, because it is a direct measurement which uses only trigonometry and requires no assumptions or physical models. On 19 December 2013 the Gaia space telescope was launched on a Soyuz rocket from French Guiana. The mission will measure, by parallax, the positions and motions of a billion stars in our galaxy over five years. This data will provide an accurate and dynamic 3D map of the galaxy, which in turn will allow for an exploration of the history of the Milky Way, because Newton's laws, which govern the motions of all these stars under the gravitational pull of each other, can be run backwards as well as forwards in time. Given precise measurements of the positions and velocities of 1 per cent of the stars in the Milky Way, it is possible to ask what the configuration of the stars looked like millions or even billions of years ago. This enables astronomers to build simulations of the evolution of our galaxy, revealing its history of collisions and mergers with other galaxies over thirteen billion years, stretching back to the beginning of the universe. Newton and Bessel would have loved it.

Stellar parallax, when deployed using a twenty-first-century orbiting observatory, is a powerful technique for mapping our galaxy out to distances of many thousands of light years. Beyond our galaxy, however, the distances are far too great to employ this direct method of distance measurement. In the mid-nineteenth century, this might have appeared an insurmountable problem, but science doesn't proceed by measurement alone. As Newton so powerfully demonstrated, scientific progress often proceeds through the interaction between theory and observation. Newton's Law of Gravitation is a theory; in physics this usually means a mathematical model that can be applied to explain or predict the behaviour of some part of the natural world. How might we measure the mass of a planet? We can't 'weigh' it directly, but given Newton's laws we can determine the planet's mass very accurately if it has a moon. The logic is quite simple – the moon's orbit clearly has something to do with the planet's gravity, which in turn has something to do with its mass. These relationships are encoded in Newton's law, and careful observation of the moon's orbit around the planet therefore allows for the planet's mass to be determined. For the more mathematical reader, the equation is:

$$M_{planet} + M_{moon} = 4\pi^2 a^3/GP^2$$

where a is the (time-averaged) distance between the planet and the moon, G is Newton's gravitational constant and P is the period of the orbit. (This equation is in fact Kepler's third law, discovered empirically by Kepler in 1619. Kepler's laws can be derived from Newton's law of gravitation.) Under the assumption that the mass of the planet is far larger than the mass of the Moon, this equation allows for the mass of the planet to be measured. This is how theoretical physics can be used to extract measurements from observation, given a mathematical model of the system. To measure the distance to objects that are too far away to use parallax, therefore, we need to find a theory or mathematical relationship that allows for a measurement of something – anything – to be related to distance. The first relationship of this type, which opened the door to all other methods of distance measurement out to the edge of the observable universe, was discovered at the end of the nineteenth century by an American astronomer named Henrietta Leavitt.

HENRIETTA LEAVITT
It was the studies of US astronomer Henrietta Leavitt (1868–1921) on the photographic magnitudes of stars that led her to discover the Cepheid variables in the Magellanic Clouds. She noticed that there was a regular variation in their brightness, and that the brighter stars had longer periods. By 1912, she had established a method which became the foundation of the one we use today to measure cosmic distances.

THE ARC IN THE SKY
Hipparcos (High Precision Parallax Collecting Satellite) was launched on 8 August 1989, but was stranded in geostationary transfer orbit by a booster failure. Despite this, the satellite has managed to measure the parallax, proper motion and position of over 120,000 stars to an accuracy of 0.002 arcseconds, about 20 times better than Earth-bound observations.

SEARCHING FOR PATTERNS IN STARLIGHT

The history of astronomy is a history of receding horizons.
Edwin Hubble

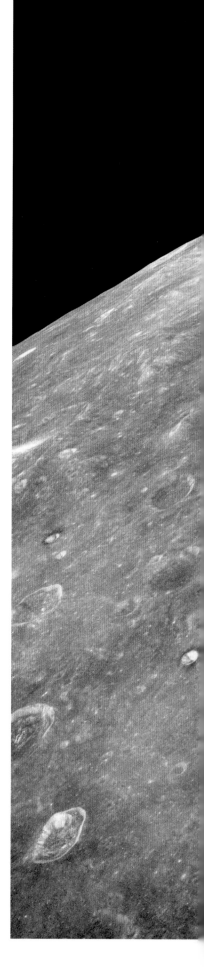

DARK SIDE OF THE MOON
This spectacular image, taken from on board the Apollo 8 spacecraft during its December 1968 mission, shows the craters that pepper the surface of the Moon. The image is taken looking towards the Southern Sea, as viewed from Earth.

LEAVITT'S LEGENDARY PAPER
Henrietta Leavitt's 1908 paper in which she noticed for the first time the relationship between the intrinsic brightness of a Cepheid variable star and the period of its variation in brightness.

The Earth is replete with features named after rogues, because history is the province of the rich and powerful and the deserving rarely become either. To find more worthy geographical nomenclature it is necessary to look further afield, to a place that escaped the attention of the vain. The dark side of the Moon is such a place, because nobody had seen it until the Soviet spacecraft Luna 3 photographed it in 1959. It isn't dark, by the way; it permanently faces away from Earth due to an effect called tidal locking, and receives the same amount of sunlight as the familiar Earth-facing side. The first humans to see it were the crew of Apollo 8, when Bill Anders memorably described it as looking like 'a sand pile my kids have played in for some time. It's all beat up, no definition, just a lot of bumps and holes.' Lacking the smooth lunar maria, the dark side is an expanse of craters, and many of these have been named entirely appropriately after deserving scientists. Giordano Bruno is there, of course, alongside Pasteur, Hertz, Millikan, D'Alembert, Planck, Pauli, Van der Waals, Poincaré, Leibnitz, Van der Graaf and Landau. Arthur Schuster, the father of the physics department at the University of Manchester, is honoured. And tucked away in the southern hemisphere, next to a plain named Apollo, is a 65-kilometre-wide partly eroded crater called Leavitt.

Henrietta Swan Leavitt was one of the 'Harvard Computers', a group of women employed to work at the Harvard College Observatory by Professor Edward Charles Pickering. By the late nineteenth century Harvard had collected a large amount of data in the form of photographic plates, but the professional astronomers had neither the time nor resources to process the reams of material. Pickering's answer was to hire women as skilled, and cheap, analysts. Scottish astronomer Williamina Fleming was his first recruit, whom he employed after proclaiming that 'even his maid' could do a better job than the overworked males at the observatory. Fleming became a respected astronomer; she was made an honorary member of the Royal Astronomical Society in London and, amongst many important published works, discovered the Horsehead Nebula in Orion. Buoyed by this successful policy, Pickering continued to expand his 'computers' throughout the later years of the nineteenth century, bringing Henrietta Leavitt into the team in 1893. Pickering assigned her to the study of stars known as variables, whose brightness changes over a period of days, weeks or months. In 1908, Leavitt published a paper based on a series of observations of variable stars in the Small Magellanic Cloud, which we now know to be a satellite galaxy of the Milky Way. It consists of a detailed list of the positions and periods of 1777 variable stars, and towards the end, a brief but extremely important observation: 'It is worthy of note that in Table VI the brighter variables have the longer periods. It is also noticeable that those having the longest periods appear to be as regular in their variations as those which pass through their changes in a day or two.'

This discovery immediately caught the interest of Pickering, and for good reason. If a star's intrinsic brightness is known, then it is a simple matter to calculate its distance. Put very simply, the further away an object is, the dimmer it appears! Leavitt and Pickering published a more detailed study in 1912, in which they proposed a simple mathematical relationship between the period and intrinsic brightness of 25 variable stars. This relationship is known as the period-luminosity relation. All that was required to calibrate the relation was a parallax measurement

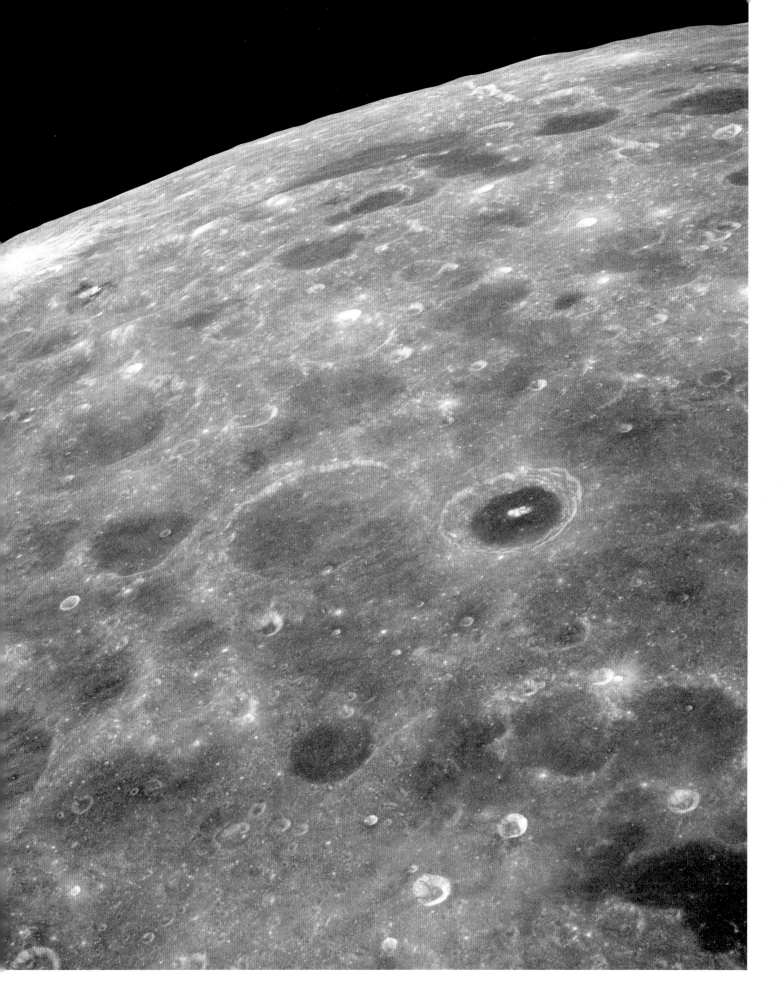

of the distance to a single variable of the type observed by Leavitt. If this could be achieved, then the distance to the Small Magellanic Cloud could be obtained. In 1913, the Danish astronomer Ejnar Hertzsprung, in a spectacularly accurate piece of astronomical observing, managed to measure the distance by parallax to the well-known variable star Delta Cephei. Delta Cephei has a period of 5.366341 days, and lies 890 light years from Earth, according to modern measurements by the Hubble Space Telescope. Because of its historic place as the first of Leavitt's variable stars to have its distance measured, these stars are now known as Cepheid variables. Inexplicably, even though Hertzsprung managed to get the parallax measurement and the distance to Delta Cephei correct, his published paper quotes the distance to the Small Magellanic Cloud as 3000 light years, which is badly wrong; the modern-day measurement is 170,000 light years. There is speculation that Hertzsprung made a simple typographical error in the paper, and for some reason couldn't be bothered to correct it. In any case, the technique had been established, and two years later Harlow Shapley published the first of a series of papers that refined the method and led him to the first measurements of the size and shape of the Milky Way. He concluded that the galaxy is a disk of stars, around 300,000 light years in extent, with the Sun positioned around 50,000 light years from the centre. This is roughly correct – the Milky Way is around 100,000 light years across and the Sun is about 25,000 light years from the centre. This was an important moment in the history of astronomy, because it was the first measurement that relegated the solar system from being the centre of everything. It's true that few if any astronomers would have claimed otherwise by the turn of the twentieth century, but science is a subject that relies on measurement rather than opinion. The journey into insignificance had begun.

BEYOND THE MILKY WAY

With the size and shape of the galaxy measured, the question of our place in creation now shifted from the position of the Sun within the galaxy to the nature of the universe itself. If the progress from Copernicus through Newton to Leavitt and Shapley appears relatively fast, certainly when viewed in the context of the glacial progress throughout the 2000-year dominance of Aristotelian thinking, then the decade that followed Shapley's determination of the size of the Milky Way might be described as an intellectual avalanche. The revolution was fuelled from two sides. A new generation of telescopes and the increasingly sophisticated observational techniques developed by astronomers like Leavitt, Hertzsprung and Shapley provided the data, and in parallel theoretical physics experienced a revolution. Claims of revolutions or paradigm shifts have to be made with great care in science – indeed the terminology is quite unfashionable in certain academic circles. But from a physicist's perspective there is no doubt that physics experienced a revolution in 1915, because in November of that year Albert Einstein presented a new theory of gravity to the Prussian Academy of Science.

The theory is known as General Relativity, and it replaces Newton's law of universal gravitation. Many physicists regard General Relativity as the most beautiful piece of physics yet devised by the human mind, and we will explore why this is so a little later. For now, let us note that the Big Bang, the expanding universe, black holes, gravitational waves and the whole evocative landscape of twenty-first-century cosmological language began, absolutely, with the publication of General Relativity. The parallels with the Newtonian revolution are clear. Without Newton's laws, there is no deep understanding of the solar system and the motions of the planets. Without General Relativity, there is no deep understanding of the large-scale structure and behaviour of the universe. But we are getting ahead of ourselves. As the second decade of the twentieth century dawned, the size and shape of the Milky Way galaxy was established, albeit with rather large errors, but the true extent of the universe beyond our galaxy was still hotly debated. Could we, at least, cling to a sort of token pre-Copernican fig leaf and place our galaxy at the centre of the universe? The desire to be special runs deep. The last intellectual rearguard action against our demotion can, rather theatrically, be said to have played out on a single evening on 26 April 1920 in the Baird auditorium at the Smithsonian Museum of Natural History, Washington DC. This is, of course, an oversimplification, but allow me a minute to enjoy the sound of the outraged shaking jowls of a thousand historians of science before I qualify and partially justify this hyperbolic claim.

CEPHEID VARIABLES

It is interesting to ask what physical processes lead to the relationship between the brightness of a Cepheid variable and its period. The details, as always, are complicated, not least because there are several different types of Cepheid, but the principle is quite simple. Delta Cephei is a yellow giant star, around 4.5 times the mass of the Sun and 2000 times more luminous. Stars of this type have large amounts of helium in their atmospheres. As the atmosphere heats up due to the nuclear fusion reactions in the core, the helium atoms are stripped of both of their electrons, forming what is called doubly ionised helium. In this form, the helium is relatively opaque, meaning that the light from the star is readily absorbed, causing the atmosphere to become hotter and to expand, increasing the stars' luminosity. But as the atmosphere expands further out into space, it cools, and ultimately the temperature drops far enough for the helium atoms to recapture an electron. This causes the gas to become transparent, allowing more light to escape. The rapidly cooling atmosphere then collapses back towards the star, heating up and restarting the cycle. The brighter the star is to begin with, the longer it takes this cycle to play out, and this is the origin of the relationship discovered by Henrietta Leavitt.

CEPHEID VARIABLES
These three images show the Cepheid variable star RZ Velorum at its minimum (left), average (centre) and maximum (right) light phases which occur as part of a cycle. The relationship between a Cepheid variable and its period was discovered by Henrietta Leavitt.

THE GREATE DEBATE

The history of science is littered with crunching moments of conflict, debates and disagreements that divided opinion in the most passionate of battles. The wonderful thing about science, however, is that the debates can be settled when facts become available. Science and 'conservative common sense' famously clashed in 1860 when Thomas Huxley and Samuel Wilberforce fulminated over the new theory of evolution published by Darwin seven months earlier. I imagine Wilberforce's indignant reddening cheeks shaking with righteous outrage as he denied the repugnant possibility that his grandfather was a monkey. None of his relatives was a chimpanzee, by the way; we simply share a common ancestor with them around 6 or 7 million years ago. But the 'unctuous, oleaginous and saponaceous' bishop, as Disraeli once called him, was having none of it. This might be a little unfair to the great Victorian orator and Bishop of the Church of England, but in the case of evolution he was firmly on the wrong side of reality. Few great leaps in knowledge occur without dividing opinion, and this is entirely appropriate. Extraordinary claims require extraordinary evidence, and the great scientific discoveries we are celebrating here are utterly extraordinary. The trick as an educated citizen of the twenty-first century is to realise that Nature is far stranger and more wonderful than human imagination, and the only appropriate response to new discoveries is to enjoy one's inevitable discomfort, take delight in being shown to be wrong and learn something as a result.

The world of astronomy had its moment of intellectual sumo in what has become known as the Great Debate. The year was 1920, and two eminent astronomers found themselves stuck on a train together travelling the 4000 kilometres from California to Washington to discuss the greatest cosmological question of the day. The younger of the two men, Harlow Shapley, we have already met. He had just published his data suggesting that the Milky Way galaxy was much larger than previously suspected. This, however, was where he believed the universe stopped; Shapley was convinced our galaxy was the beginning and end of the cosmos. His travelling companion thought otherwise. Heber Curtis had been studying a misty patch of light known as the Andromeda Nebula. He was convinced that this was not part of our galaxy, but a separate island universe of billions of other stars.

It is not known what they discussed on the train, but the debate itself took place at the Smithsonian Museum of Natural History throughout the day and night of 26 April. At stake was the scale of the universe itself, and both men knew that the question would ultimately be settled by evidence rather than debating skills. The human race had already been shunted from the centre of the universe by Copernicus, and now faced the possibility that the Milky Way galaxy itself was part of a multitude, stretching across millions of light years of space. The question wasn't settled that evening, but the experienced Curtis, perceived as the underdog because of the magnitude of what he was suggesting, landed significant blows. Curtis observed that the Andromeda Nebula contains a number of novae – exploding stars that shine temporarily, but brightly, in the night sky – but he also noted that the novae in Andromeda appeared on average to be ten times fainter than any others. Curtis asserted that Andromeda's novae appear dimmer simply because they are perhaps half a million light years further away than those in the Milky Way. Andromeda is therefore another galaxy, claimed Curtis, which strongly implied that the other so-called nebulae were other galaxies too. This was the very definition of an extraordinary claim, and the extraordinary evidence came only four years later.

HUBBLE AND HOOKER
This, the 100-inch Hooker telescope at Mount Wilson Observatory, California, was the instrument that enabled Edwin Hubble (right) to observe the Cepheid variable stars in the Andromeda Nebula, and by analysis of his findings prove that the galaxy was located outside our own. Hubble is pictured with the then director of the Observatory, Walter Adams (centre), and British astronomer James Jeans (left).

In 1923 a photo of Andromeda, taken by a 33-year-old astronomer called Edwin Hubble, further fuelled the Great Debate. It's only a photograph but, just like Anders' Earthrise, it belongs to a rarefied group of images that have transformed our perspective. Aside from their scientific merit, such images assume great cultural significance because of the ideas they generate and the philosophical and ideological challenges they pose. They also carry with them, in the shadows, personal stories. Someone would have taken a photograph of Andromeda, someday, and discovered what Hubble did. But Hubble took this one, and his story therefore becomes inextricably intertwined with it. Some don't like their history presented in this way, but science is richer when its stories include people as well as ideas; curiosity is, after all, a human virtue. Hubble may never have taken the photograph had he followed through on a promise to his father to practise law. Reading jurisprudence at Queen's College, Oxford, as one of the first Rhodes Scholars, Hubble aimed to fulfil his father's wishes, but John Hubble died before Edwin finished his degree. The death of his father encouraged Edwin to ditch law and revisit his childhood passion for astronomy. He left Oxford for the University of Chicago, joined the Yerkes Observatory and received his PhD in 1917 with a thesis entitled 'Photographic Investigations of Faint Nebulae'. After brief service in the US Army at the end of World War One, Hubble obtained a position at the Mount Wilson Observatory. He found himself at the controls of the largest, most powerful telescope on the planet, and with the knowledge and good sense to point it at the most intriguing and controversial object in the night sky: Andromeda. Just like Curtis before him, Hubble could make out distinct features within the misty patch, but the newly commissioned 100-inch Hooker telescope allowed him to see much more detail. On 5 October 1923 he took a 45-minute exposure, found three unidentified specks that he assumed were new novae and marked them all with an 'N'.

ANDROMEDA
Photographs of Andromeda like this one have transformed our perspective. Such images question long-held beliefs and theories and open up the debate about our history and place in the universe.

To confirm his findings Hubble needed to compare this plate with previous images of Andromeda taken at Mount Wilson. The following day he made the journey down to the basement archive where the observatory's collection of images was catalogued and stored. To Hubble's delight, two of the specks were indeed newly discovered novae – what we now know to be the bright nuclear flares of white dwarf stars as they accrete gas and dust from a nearby companion. But it was the third speck that he found most interesting when he compared it to previous images. As Hubble scanned back through the Mount Wilson catalogue he discovered that the star had been captured before; in some plates it appeared brighter, whereas in others it appeared dim or not present at all. Hubble immediately grasped the importance of his discovery. The third speck was a Cepheid variable, the type of star Henrietta Leavitt had studied two decades earlier. In one of the most famous corrections in scientific history, Hubble crossed out the letter 'N' and replaced it in red ink with the letters 'VAR' followed by a very understated exclamation mark.

Hubble had discovered a cosmic yardstick in Andromeda, and it was a trivial matter to calculate the distance. The new star varied with a period of 31.415 days, which, following Leavitt, implied its intrinsic brightness was 7000 times that of our Sun, and yet it appeared so dim in the night sky that it was invisible to all but the most powerful of telescopes. Hubble's initial calculations revealed that the star was over 900,000 light years away from Earth, a staggering distance when the size of our own galaxy was estimated to be no more than 100,000 light years across. Hubble, with the help of Leavitt's ruler, laid the Great Debate to rest. Andromeda, the distant patch of light in the night sky, is a galaxy; an island, according to current estimates, of a trillion suns. Current measurements put the giant spiral at a distance of 2.5 million light years from the Milky Way, one of around 54 galaxies gravitationally bound together to form our galactic neighbourhood known as The Local Group.

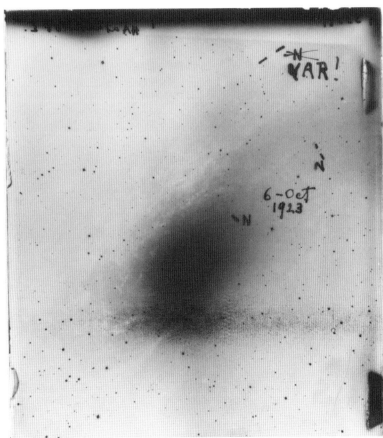

HUBBLE'S EUREKA MOMENT
Edwin Hubble's glass plate from the Hooker telescope very clearly reveals his excitement at his discovery that one of the novae he thought he had previously located was in fact a variable – VAR! marks the spot.

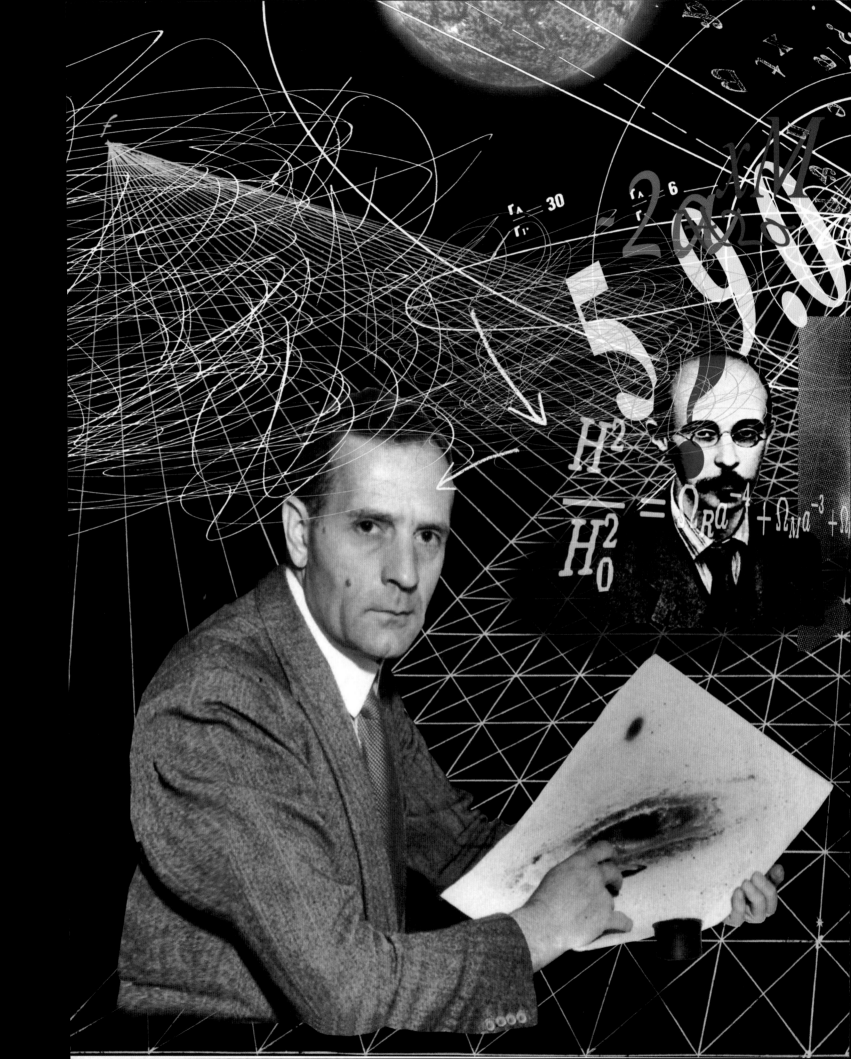

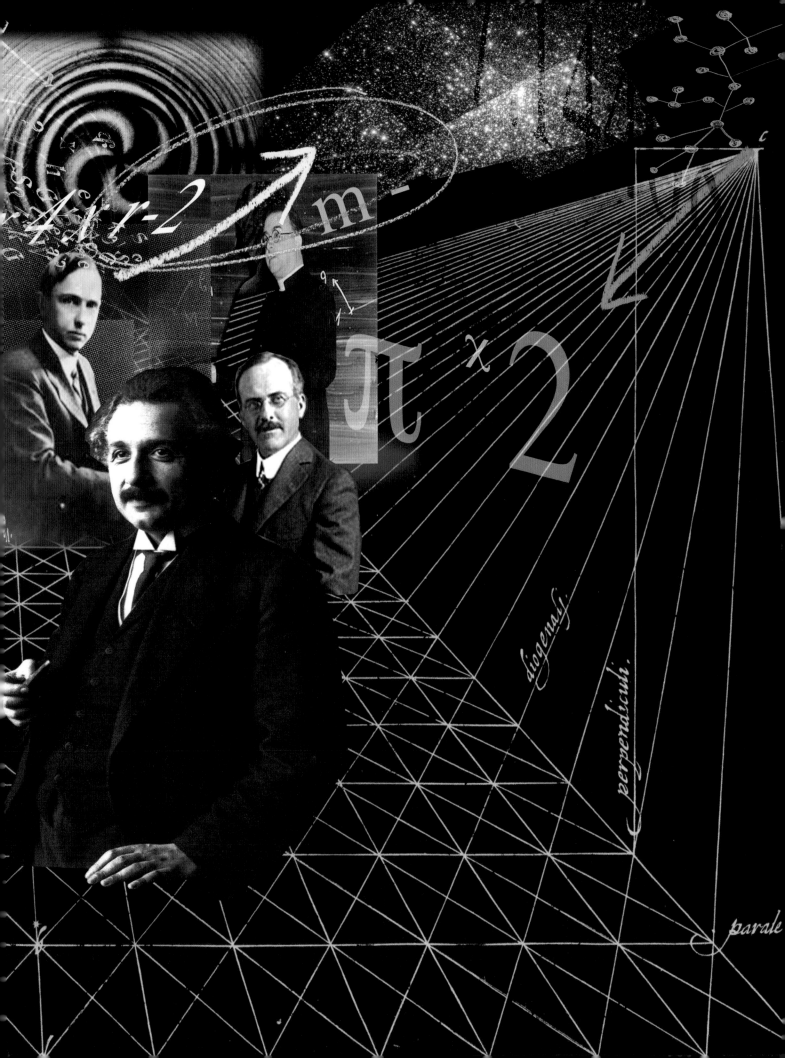

THE POLITICAL RAMIFICATIONS OF REALITY, OR 'HOW TO AVOID GETTING LOCKED UP'

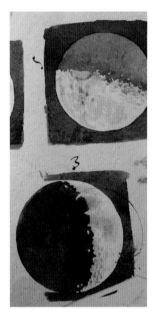

FIRST VIEWS OF THE MOON
Galileo's beautiful watercolours of the Moon, drawn in around November–December 1609, are famous as the first realistic depictions of the Moon viewed through a telescope from Earth, and were revolutionary at the time for suggesting that the lunar surface was not perfectly smooth.

THE MADONNA ON THE MOON
High up in the dome above the Pauline Chapel in the church of Santa Maria Maggiore in Rome, a spectacular fresco depicts the Madonna on a lunar landscape that is pitted and uneven. In his painting, Lodovico Cardi – better known as Cigoli – reveals the influence of his friend Galileo and his discoveries.

What is science? There are many answers, and whole academic careers are devoted to a complex analysis of the historical and sociological development of the subject. To a working scientist, however, I think the answer is quite simple and illuminating because it reveals a lot about how scientists see themselves and what they do. The great (an overused adjective, but not in this case) physicist Richard Feynman gave a characteristically clear and simple description in his Messenger Lectures delivered at Cornell University in 1964: 'In general, we look for a new law by the following process: First we guess it. Then we – now don't laugh, that's really true – then we compute the consequences of the guess to see what, if this is right, if this law that we guessed is right, to see what it would imply. And then we compare the computation results to nature, or we say compare to experiment or experience, compare it directly with observations to see if it works. If it disagrees with experiment, it's wrong. In that simple statement is the key to science. It doesn't make any difference how beautiful your guess is, it doesn't make any difference how smart you are, who made the guess, or what his name is. If it disagrees with experiment, it's wrong. That's all there is to it.'

Why do I like this so much? The reason is that it is modest – almost humble in its simplicity – and this, in my opinion, is the key to the success of science. Science isn't a grandiose practice; there are no great ambitions to understand why we are here or how the whole universe works or our place within it, or even how the universe began. Just have a look at something – the smallest, most trivial little thing – and enjoy trying to figure out how it works. That is science. In a famous BBC *Horizon* film broadcast in 1982 called 'The Pleasure of Finding Things Out', Feynman went further: 'People say to me, "Are you looking for the ultimate laws of physics?" No, I'm not. I'm just looking to find out more about the world and if it turns out there is a simple ultimate law which explains everything, so be it; that would be very nice to discover. If it turns out it's like an onion with millions of layers and we're just sick and tired of looking at the layers, then that's the way it is…. My interest in science is to simply find out more about the world.'

The remarkable thing about science, however, is that it has ended up addressing some of the great philosophical questions about the origin and fate of the universe and the meaning of existence without actually setting out to do so, and this is no accident. You won't discover anything meaningful about the world by sitting on a pillar for decades and contemplating the cosmos, although you may become a saint. No, a truly deep and profound understanding of the natural world has emerged more often than not from the consideration of much less lofty and profound questions, and there are two reasons for this. Firstly, simple questions can be answered systematically by applying the scientific method as outlined by Richard Feynman, whereas complex and badly posed questions such as 'Why are we here?' cannot. But more importantly, and rather more profoundly, it turns out that the answers to simple questions can overturn centuries of philosophical and theological pontificating quite by accident. Reputations count for naught in the face of observation. The famous story of Galileo's clashes with the Inquisition at the height of the Copernican debate, which he certainly did not expect (nobody does), is the archetypal example.

Galileo began his university career with the study of medicine, but his imagination was captured by art and mathematics. Between studying Medicine in Pisa and returning to his hometown in 1589 to

become Professor of Mathematics, Galileo spent a year in Florence teaching perspective and in particular a technique called chiaroscuro. Chiaroscuro is the study of light and shadow, and how it can be used to create a sense of depth by accurately representing the way that light sources illuminate objects. Chiaroscuro was one of the most important new artistic techniques to emerge during Galileo's time, allowing a new sense of realism to be portrayed on canvas.

Although Galileo spent only a brief time in Florence, the skills he acquired had a great impact on his scientific work. In particular, his carefully developed ability to understand the delicate play of light on three-dimensional shapes, when applied to his later astronomical studies, played an important role in undermining the Aristotelian cosmological edifice which formed a cornerstone of the teachings of the Roman Catholic Church.

The small and seemingly innocuous theological thread on which Galileo unwittingly tugged was made available to him on a visit to Venice in 1609, when he purchased the lenses required to build his first telescope. One of the first objects he turned his 'perspective tube' towards was the Moon. With the mind of a mathematician and the eye of an artist, Galileo drew a series of six watercolours representing what he saw.

These images are both beautiful and revolutionary. Catholic dogma asserted that the Moon and the other heavenly bodies were perfect, unblemished spheres. Previous astronomers who had viewed the Moon, either with the naked eye or through telescopes, had drawn a two-dimensional blotchy surface, but Galileo saw the patterns of light and dark differently. His training in chiaroscuro revealed to him an alien lunar landscape of mountain ranges and craters.

'I have been led to the conclusion that ... the surface of the Moon is not smooth, even and perfectly spherical – as the great crowd of philosophers have believed about this and other heavenly bodies – but, on the contrary, to be uneven and rough and crowded with depression and bulges. And it is like the face of the Earth itself, which is marked here and there with chains of mountains and depths of valleys.'

Galileo shared the watercolours with his long-standing friend from Florence, the artist Cigoli, who was inspired to represent this new and radical view of the Moon in the grandest of settings. Built in the year 430 by Pope Sixtus III, the Pauline Chapel in Rome documents the changing artistic styles and techniques used to represent the natural world across many centuries; a place filled with shifting examples of how the three-dimensional world can be represented on a two-dimensional surface. Covering the dome of the Pauline Chapel is Cigoli's final masterpiece – a striking fresco depicting a familiar scene of the Virgin Mary bathed in a shaft of golden light surrounded by cherubs and angels. The fresco depicts Mary over what was, for the first time, a detailed, textured and cratered moon. The Vatican named it the Assumption of the Virgin, unaware perhaps of the philosophical challenge it represented. Here was art representing scientific knowledge – a type of knowledge radically different to historical or scriptural authority, based on observation rather than dogma and presented unashamedly in a grand setting for all in Rome to see. It is undoubtedly true that Galileo didn't intend to challenge the very theological foundations of the Church of Rome by observing the Moon through a telescope. But scientific discoveries, however innocuous they may seem at first sight, have a way of undermining those who don't much care for facts. Reality catches up with everyone eventually.

With his depictions of the Moon completed, Galileo turned his ever more powerful lenses to other celestial bodies. Between 7 and 13 January 1610, he became the first human to observe Jupiter's four largest moons – Io, Europa, Ganymede and Callisto – now known as the Galilean

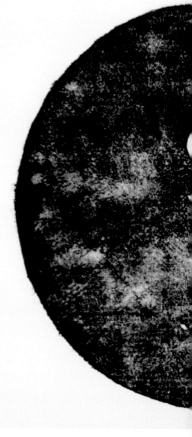

GALILEO'S MOON
This sketch of the Moon is one of several made by Galileo in 1610 from observations he made through a telescope he himself had built two years previously.

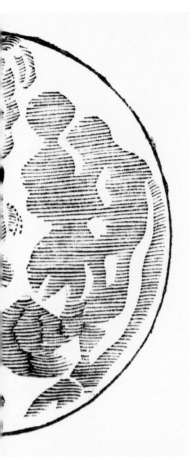

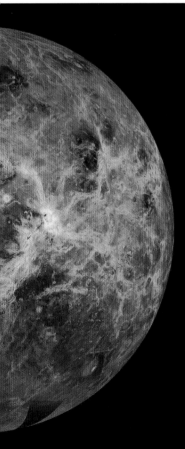

Satellites. For Galileo, this was further evidence to support the work of Copernicus and the physical reality of the heliocentric model. If moons were orbiting Jupiter, Galileo reasoned, it was impossible to argue that the Earth was at the centre of the universe, because heavenly bodies existed that did not circle the Earth.

Galileo published these observations in the spring of 1610 in 'The Starry Messenger', and from his correspondence with Kepler his irritation with the discontent it caused amongst philosophers was clear. 'My dear Kepler, I wish that we might laugh at the remarkable stupidity of the common herd. What do you have to say about the principal philosophers of this academy who are filled with the stubbornness of an asp and do not want to look at either the planets, the Moon or the telescope, even though I have freely and deliberately offered them the opportunity a thousand times? Truly, just as the asp stops its ears, so do these philosophers shut their eyes to the light of truth.'

To Galileo's mind, absolute confirmation of Copernicus's heliocentric model was provided by his studies of Venus. Beginning in September 1610, Galileo observed Venus over the course of months and, like the Moon, he observed that Venus had phases. Sometimes the planet was lit completely by the Sun, but at other times only a crescent appeared to be illuminated. The only plausible explanation for this observation was that Venus was orbiting the Sun. This was surely final compelling evidence of a solar system with the Sun at its heart and the planets orbiting around it.

It wasn't that simple, of course. Galileo, in what was certainly an ill-judged move, decided to move beyond reporting his scientific observations and instead champion a particular theological and philosophical interpretation of the data – namely that the Church was wrong and that the Earth was most definitely not the centre of the universe. This he seems to have done because he wanted to be famous, and famous he became. Copernicus's *De revolutionibus* was banned until 'corrected' (the full version was not removed from the banned list until 1758!) and Galileo ordered not to repeat his 'foolish and absurd' conclusions. Galileo didn't keep quiet, and he achieved his historical notoriety by being put under house arrest in 1633, where he stayed for the remainder of his life.

Many historians characterise Galileo as a bit of an egotistical social climber who brought it all on himself, which is partly true and yet also desperately unfair. He was undoubtedly a great scientist and a supremely talented astronomical observer. In particular, he was the first to clearly state the principle of relativity which lies at the heart of Newton's laws of motion; namely that there is no such thing as absolute rest or absolute motion. This is why we don't feel the movement of the Earth around the Sun, and why Aristotle et al. were misled into reading far too much into their stationary feelings. In the hands of Albert Einstein, the principle of relativity can be generalised to freely falling objects in a gravitational field, and this ultimately leads to modern cosmology and the Big Bang theory. But we are jumping ahead again. The purpose of recounting the story of Galileo is not to attack the easy target of the Inquisition (which nobody expects). Rather, it is to highlight the fact that the smallest and most modest of scientific observations can lead to great philosophical and theological shifts that in turn can have a tremendous impact on society. Galileo, by looking through a telescope, doing some drawings and thinking about what he saw, helped to undermine centuries of autocratic idiocy and woolly thinking. In doing so, he got himself locked up, but also bridged the gap between Copernicus and Kepler, and paved the way for Isaac Newton and ultimately Albert Einstein to construct a complete description of the universe and our place within it.

VENUS
Galileo's observations were not just restricted to the Moon, for he also studied Venus. This false-colour projection shows the western hemisphere of Venus with the North Pole at its apex.

THE HAPPIEST THOUGHT
OF MY LIFE

Scientific progress, then, is often triggered by rather innocuous discoveries or simple realisations. There is a terrible cliché about scientists exhibiting a 'childlike' fascination with nature, but I can't think of a better way of putting it. The sense in which the cliché rings true is that children are occasionally in the habit of focusing on a very small thing and continuing to ask the question 'Why?' until they get an answer that satisfies their curiosity. Adults don't seem to do this as much. Good scientists do, however, and if I have a thesis in this chapter then it is as follows; by focusing on tiny but interesting things with honesty and clarity, great and profound discoveries are made, often by flawed human beings who don't initially realise the consequences of their investigations. The absolutely archetypal example of such an approach can be found at the beginning of Einstein's quest to replace Newton's Theory of Gravity.

Einstein is most famous for his equation $E=mc^2$, which is contained within the special theory of relativity he published in 1905. At the heart of the theory is a very simple concept that dates all the way back to Galileo. Put simply, there is no way that you can tell whether you are moving or not. This sounds a bit abstract, but we all know it's true. If you are sitting in a room at home reading this book, then it feels the same as if you are sitting in an aircraft reading this book, as long as there is no turbulence and the aircraft is in level flight. If you aren't allowed to look out of the window, then nothing you can do in the room or on the plane will tell you whether or not you are 'sitting still' or moving. You might claim that your room is self-evidently not moving, whereas a plane obviously is because otherwise it wouldn't take you from London to New York. But that's not right, because your room is moving in orbit around the Sun, and indeed it is spinning around the Earth's axis, and the Sun itself is in orbit around the galaxy, which is moving relative to other galaxies in the universe. Einstein discovered his famous equation $E=mc^2$ by taking this seemingly pedantic reasoning seriously and asserting that NO experiment you can ever do, even in principle, using clocks, radioactive atoms, electrical circuits, pendulums, or any physical object at all, will tell you whether or not you are moving. Anyone has the absolute right to claim that they are at rest, as long as there is no net force acting on them causing them to accelerate. You are claiming it now, no doubt, if you are reading this book sitting comfortably on your sofa. Pedantry is very useful sometimes, because without Einstein's theory of special relativity we wouldn't have $E=mc^2$, we wouldn't really understand nuclear or particle physics, how the Sun shines or how radioactivity works. We wouldn't understand the universe.

Something important bothered Einstein after he published his theory in 1905, however. Newton's great achievement – the all-conquering Universal Law of Gravitation – did not fit within the framework of special relativity, and therefore one or the other required modification. Einstein's response to this problem was typically Einsteinian; he thought about it very carefully, and, in November 1907, whilst sitting in his chair in the patent office in Bern, he found the right thread to pull. Looking back at the moment in an article written in 1920, Einstein described his idea with beautiful, and indeed child-like, simplicity.

EINSTEIN'S BEAUTIFUL THEORY
Albert Einstein, one of the great minds of science, whose General Theory of Relativity, published in 1916, is often cited as the most beautiful scientific theory of all.

'Then there occurred to me the "*glücklichste Gedanke meines Lebens*", the happiest thought of my life, in the following form. The gravitational field has only a relative existence in a way similar to the electric field generated by magnetoelectric induction. *Because for an observer falling freely from the roof of a house there exists* – at least in his immediate surroundings – *no gravitational field* [his italics]. Indeed, if the observer drops some bodies then these remain relative to him in a state of rest or of uniform motion, independent of their particular chemical or physical nature (in this consideration the air resistance is, of course, ignored). The observer therefore has a right to interpret his state as "at rest".'

I am well aware that you might object quite strongly to this statement, because it appears to violate common sense. Surely an object falling under the action of the gravitational force is accelerating towards the ground, and therefore cannot be said to be 'at rest'? Good, because if you think that then you are about to learn a valuable lesson. Common sense is completely worthless and irrelevant when trying to understand reality. This is probably why people who like to boast about their common sense tend to rail against the fact that they share a common ancestor with a monkey. How, then, to convince you that Einstein was, and indeed still is, correct?

Most of the time, books are better at conveying complex ideas than television. There are many reasons for this, some of which I'll discuss in a future autobiography when my time on TV is long over. But when done well, television pictures can convey ideas with an elegance and economy unavailable in print. *Human Universe* contains, I hope, some of these moments, but there is one sequence in particular that I think fits into this category.

NASA's Plum Brook Station in Ohio is home to the world's largest vacuum chamber. It is 30 metres in diameter and 37 metres high, and was designed in the 1960s to test nuclear rockets in simulated space-like conditions. No nuclear rocket has ever been fired inside – the programme was cancelled before the facility was completed – but many spacecraft, from the Skylab nosecone to the airbags on Mars landers, have been tested inside this cathedral of aluminium. To my absolute delight, NASA agreed to conduct an experiment using their vacuum chamber to demonstrate precisely what motivated Einstein to his remarkable conclusion. The experiment involves pumping all the air out of the chamber and dropping a bunch of feathers and a bowling ball from a crane. Both Galileo and Newton knew the result, which is not in question. The feathers and the bowling ball both hit the ground at the same time. Newton's explanation for this striking result is as follows. The gravitational force acting on a feather is proportional to its mass. We've already seen this written down in Newton's Law of Gravitation. That gravitational force causes the feather to accelerate, according to Newton's other equation, $F=ma$. This equation says that the more massive something is, the more force has to be applied to make it accelerate. Magically, the mass that appears in $F=ma$ is precisely the same as the mass that appears in the Law of Gravitation, and so they precisely cancel each other out. In other words, the more massive something is, the stronger the gravitational force between it and the Earth, but the more massive it is, the larger this force has to be to get it moving. Everything cancels out, and so everything ends up falling at the same rate. The problem with this explanation is that nobody has ever thought of a good reason why these two masses should be the same. In physics, this is known as the equivalence principle, because 'gravitational mass' and 'inertial mass' are precisely equivalent to each other.

Einstein's explanation for the fact that both the feathers and the bowling ball fall at the same rate in the Plum Brook vacuum chamber is radically different. Recall Einstein's happiest thought. 'Because for an observer falling freely from the roof of a house there exists ... no gravitational field'. There is no force acting on the feathers or the ball in freefall, and therefore they don't accelerate! They stay precisely where they are; at rest, relative to each other. Or, if you prefer, they stand still because we are always able to define ourselves as being at rest if there are no forces acting on us. But, you are surely asking, how come they eventually hit the ground if they are not moving because no forces are

SCIENCE IN ACTION
Plum Brook Station, part of NASA's Glenn Research Center, is home to the world's largest space simulator – the world's biggest vacuum chamber – which we used to demonstrate's Einstein's theory of general relativity.

acting on them? The answer, according to Einstein, is that the ground is accelerating up to meet them, and hits them like a cricket bat! But, but, but, you must be thinking, I'm sitting on the ground now and I'm not accelerating. Oh yes you are, and you know it because you can feel a force acting on you. It's the force exerted by the chair on which you may be sitting, or the ground on which you are standing. This is obvious – if you stand up long enough then your feet will hurt because there is a force acting on them. And if there is a force acting on them, then they are accelerating. There is no sleight of hand here. The very beautiful thing about Einstein's happiest thought is that, once you know it, it's utterly obvious. Standing on the ground is hard work because it exerts a force on you. The effect is precisely the same as sitting in an accelerating car and being pushed back into your seat. You can feel the acceleration viscerally, and if you switch off your common sense for a moment, then you can feel

the acceleration now. The only way you can get rid of the acceleration, momentarily, is to jump off a roof.

This is wonderful reasoning, but of course it does raise the thorny question of why, if there is no such thing as gravity, the Earth orbits the Sun. Maybe Aristotle was right after all. The answer is not easy, and it took Einstein almost a decade to work out the details. The result, published in 1916, is the General Theory of Relativity, which is often cited as the most beautiful scientific theory of them all. General Relativity is notoriously mathematically and conceptually difficult when you get into the details of making predictions that can be compared with observations. Indeed, most physics students in the UK will not meet General Relativity until their final year, or until they become postgraduates. But having said that, the basic idea is very simple. Einstein replaced the force of gravity with geometry – in particular, the curvature of space and time.

Imagine that you are standing on the surface of the Earth at the equator with a friend. You both start walking due north, parallel to each other. As you get closer to the North Pole, you will find that you move closer together, and if you carry on all the way to the Pole you will bump into each other. If you don't know any better, then you may conclude that there is some kind of force, pulling you both together. But in reality there is no such force. Instead, the surface of the Earth is curved into a sphere, and on a sphere, lines that are parallel at the equator meet at the Pole – they are called lines of longitude. This is how geometry can lead to the appearance of a force.

Einstein's theory of gravity contains equations that allow us to calculate how space and time are curved by the presence of matter and energy and how objects move across the curved spacetime – just like you and your friend moving across the surface of the Earth. Spacetime is often described as the fabric of the universe, which isn't a bad term. Massive objects such as stars and planets tell the fabric how to curve, and the fabric tells objects how to move. In particular, all objects follow 'straight line' paths across the curved spacetime that are known in the jargon as geodesics. This is the General Relativistic equivalent of Newton's first law of motion – every body continues in a state of rest or uniform motion in a straight line unless acted upon by a force. Einstein's description of the Earth's orbit around the Sun is therefore quite simple. The orbit is a straight line in spacetime curved by the presence of the Sun, and the Earth follows this straight line because there are no forces acting on it to make it do otherwise. This is the opposite of the Newtonian description, which says that the Earth would fly through space in what we would intuitively call a 'straight line' if it were not for the force of gravity acting between it and the Sun. Straight lines in curved spacetime look curved to us for precisely the same reason that lines of longitude on the surface of the Earth look curved to us; the space upon which the straight lines are defined is curved.

This is all well and good, but there may be a question that has been nagging away in your mind since I told you that the ground accelerated up and hit the feathers and the bowling ball at Plum Brook like a cricket bat. How could it possibly be that every piece of the Earth's surface is accelerating away from its centre, and yet the Earth stays intact as a sphere with a fixed radius? The answer is that if a little piece of the Earth's surface at Plum Brook were left to its own devices, it would do precisely the same thing as the feather and the bowling ball; it would follow a straight line through spacetime. These straight lines point radially inwards towards the centre of the Earth. This is the 'state of rest', if you like – the natural trajectory that would be followed by anything. The geodesics point radially inwards because of the way that the mass of the Earth curves spacetime. So a collapsing Earth would be the natural state of things without any forces acting – one in which, ultimately, all the matter

10:35:38:09

TESTING EINSTEIN'S THEORY
In the great vaccum chamber at Plum Brook Station we re-created Galileo's simple experiment by dropping a heavy object (bowling ball) and some lighter ones (feathers) to see which falls faster.

10:35:49:03

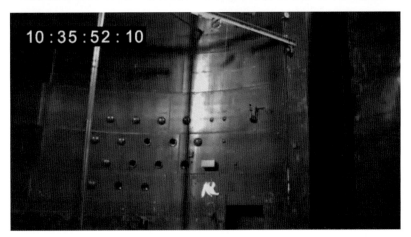

10:35:52:10

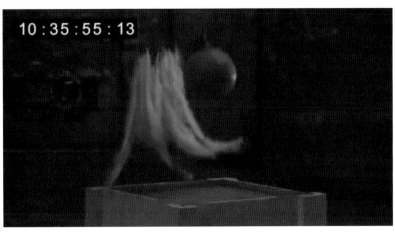

10:35:55:13

would collapse into a little black hole. The thing that prevents this from happening is the rigidity of the matter that makes up the Earth, which ultimately has its origin in the force of electromagnetism and a quantum mechanical effect called the Pauli Exclusion Principle. In order to stay as a big, spherical, Earth-sized ball, a force must act on each little piece of ground and this must cause each piece of ground to accelerate. Every piece of big spherical things like planets must continually accelerate radially outwards to stay as they are, according to General Relativity.

From what I've said so far, it might seem that General Relativity is simply a pleasing way of explaining why the Earth orbits the Sun and why objects all fall at the same rate in a gravitational field. General Relativity is far more than that, however. Very importantly, it makes precise predictions about the behaviour of certain astronomical objects that are radically different from Newton's. One of the most spectacular examples is a binary star system known rather less than poetically as PSR J0348+0432. The two stars in this system are exotic astrophysical objects. One is a white dwarf, the core of a dead star held up against the force of gravity by a sea of electrons. Electrons behave according to the Pauli Exclusion Principle, which, roughly speaking, states that electrons resist being squashed together. This purely quantum mechanical effect can halt the collapse of a star at the end of its life, leaving a super-dense blob of matter. White dwarfs are typically between 0.6 and 1.4 times the mass of our Sun, but with a volume comparable to that of the Earth. The upper limit of the mass of a white dwarf is known as the Chandrasekhar limit, and was first calculated by the Indian

astrophysicist Subrahmanyan Chandrasekhar in 1930. The calculation is a tour de force of modern physics, and relates the maximum mass of these exotic objects to four fundamental constants of nature – Newton's Gravitational constant, Planck's constant, the speed of light and the mass of the proton. After almost a century of astronomical observations, no white dwarf has ever been discovered that exceeds the Chandrasekhar limit. Almost all the stars in the Milky Way, including our Sun, will end their lives as white dwarfs. Only the most massive stars will produce a remnant that exceeds the Chandrasekhar limit, and the vast majority of these will produce an even more exotic object known as a neutron star. In the PSR J0348+0432 system, quite wonderfully, the white dwarf has a neutron star companion, and this is what makes the system so special.

If the remains of a star exceed the Chandrasekhar limit, the electrons are squashed so tightly onto the protons in the star that they can react together via the weak nuclear force to produce neutrons (with the emission of a particle called a neutrino). Through this mechanism, the whole star is converted into a giant atomic nucleus. Neutrons, just like electrons, obey the Pauli Exclusion Principle and resist being squashed together, leading to a stable dead star. Neutron stars can have masses several times that of our Sun, but quite astonishingly are only around 10 kilometres in diameter. They are the densest stars known; a teaspoon-full of neutron star matter weighs as much as a mountain.

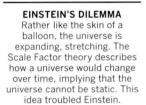

EINSTEIN'S DILEMMA
Rather like the skin of a balloon, the universe is expanding, stretching. The Scale Factor theory describes how a universe would change over time, implying that the universe cannot be static. This idea troubled Einstein.

Imagine, for a moment, this exotic star system. The white dwarf and neutron star are very close together; they orbit around each other at a distance of 830,000 kilometres – that's around twice the distance to the Moon – once every 2 hours and 27 minutes. That's an orbital velocity of around 2 million kilometres per hour. The neutron star is twice the mass of our Sun, around 10 kilometres in diameter, and spins on its axis 25 times a second. This is a star system of unbelievable violence. Einstein's Theory of General Relativity predicts that the two stars should spiral in towards each other because they lose energy by disturbing spacetime itself, emitting what are known as gravitational waves. The loss of energy is minuscule, resulting in a change in orbital period of eight millionths of a second per year. In a triumph of observational astronomy, using the giant Arecibo radio telescope in Puerto Rico, the Effelsberg telescope in Germany and the European Southern Observatory's VLT in Chile, astronomers measured the rate of orbital decay of PSR J0348+0432 in 2013 and found it to be precisely as Einstein predicted. This is quite remarkable. Einstein could never have dreamt of the existence of white dwarfs and neutron stars when he had his happiest thought in 1907, and yet by thinking carefully about falling off a roof he was able to construct a theory of gravity that describes, with absolute precision, the behaviour of the most exotic star system accessible to twenty-first-century telescopes. And that, if I really need to say it, is why I love physics.

Einstein's Theory of General Relativity has, at the time of writing, passed every precision test that scientists have been able to carry out in the century since it was first published. From the motion of feathers and bowling balls in the Earth's gravitational field to the extreme astrophysical violence of PSR J0348+0432, the theory comes through with flying colours.

There is rather more to Einstein's magisterial theory than the mere description of orbits, however. General Relativity is fundamentally different to Newton's theory because it doesn't simply provide a model for the action of gravity. Rather, it provides an explanation for the existence of the gravitational force itself in terms of the curvature of spacetime. It's worth writing down Einstein's field equations, because they are (to be honest) deceptively simple.

$$G_{\mu\nu} = 8\pi G T_{\mu\nu}$$

**ANCIENT IDEAS, MODERN
TECHNOLOGY**

In 2013 astronomers at the
Arecibo observatory in Puerto
Rico confirmed what Einstein
had purported in 1907. Using
their radio telescope – the
biggest single dish in the
world – they found that the
rate of orbital decay of PSR
J0348+0432 was exactly as
Einstein had predicted over 100
years earlier.

Here, the right-hand side describes the distribution of matter and energy
in some region of spacetime, and left-hand side describes the shape of
spacetime as a result of the matter and energy distribution. To calculate
the orbit of the Earth around the Sun one would put a spherical distribution
of mass with the radius of the Sun into the right-hand side of the equation,
and (roughly speaking) out would pop the shape of spacetime around the
Sun. Given the shape of spacetime, the orbit of the Earth can be calculated.
It's not completely trivial to do this by any means, and the notation above
hides great complexity. But the point is simply that, given some distribution
of matter and energy, Einstein's equations let you calculate what spacetime
looks like. But here is the remarkable point that draws us towards the end
of our story. Einstein's equations deal with the shape of spacetime – the
fabric of the universe. The first thing to note is that we are dealing with
spacetime, not just space. Space is not a fixed arena within which things
happen with a big universal clock marking some sort of cosmic time upon
which everyone agrees. The fabric of the universe in Einstein's theory is a
dynamical thing. Very importantly, therefore, Einstein's equations don't
necessarily describe something that is static and unchanging. The second
thing to note is that nowhere have we restricted the domain of Einstein's
theory to the region of spacetime around a single star, or even a double
star system such as PSR J0348+0432. Indeed, there is no suggestion in
Einstein's theory that such a restriction is necessary. Einstein's equations
can be applied to an unlimited region of spacetime. This implies that
they can, at least in principle, be used to describe the shape and evolution
of the entire universe.

A DAY WITHOUT YESTERDAY

There were two ways of arriving at the truth;
I decided to follow them both.
Georges Lemaître

Storytelling is an ancient and deeply embedded human impulse; we learn, we communicate, we connect across generations through stories. We use them to explore the minutiae of human life, taking delight in the smallest things. And we tell grander tales of origins and endings. History is littered with stories about the creation of the universe; they seem as old as humanity itself. Multifarious gods, cosmic eggs, worlds emerging from chaos or order, from the waters or the sky or nothing at all – there exist as many creation myths as there are cultures. The impulse to understand the origin of the universe is clearly a powerful unifying idea, although the very existence of many different mythologies continues to be a source of division. It is an unfortunate testament to the emotional power of creation narratives that so much energy is spent arguing about old ones rather than using the increasingly detailed observational evidence available to twenty-first-century citizens to construct new ones. We live in a very privileged and exciting time in this sense, because observational evidence for creation stories was scant even a single lifetime ago. When my grandparents were born in Oldham at the turn of the twentieth century, there was no scientific creation story. Astronomers were not even aware of a universe beyond the Milky Way, which makes it all the more remarkable that the modern scientific approach to the description of the universe emerged almost fully formed from Einstein's Theory of General Relativity before Edwin Hubble published the discovery of his Cepheid variable star in Andromeda and settled Shapley and Curtis's Great Debate.

One of the beautiful things about mathematical physics is that equations contain stories. If you think of equations in terms of the nasty little things you used to solve at school on a damp autumn afternoon, then that may sound like a strange and abstract idea. But equations like Einstein's field equations are much more complex animals. Recall that Einstein's equations will tell you the shape of spacetime, given some distribution of matter and energy. That shape is known as a solution of the equations, and it is these solutions that contain the stories. The first exact solution to Einstein's field equations was discovered in 1915 by the German physicist Karl Schwarzschild. Schwarzschild used the equations to calculate the shape of spacetime around a perfectly spherical, non-rotating mass. Schwarzschild's solution can be used to describe planetary orbits around a star, but it also contains some of the most exotic ideas in modern physics; it describes what we now know as the event horizon of a black hole. The well-known tales of astronauts being spaghettified as they fall towards oblivion inside a supermassive collapsed star are to be found in Schwarzschild's solution. The calculation was a remarkable achievement, not least because Schwarzschild completed it whilst serving in the German Army at the Russian Front. Shortly afterwards, the 42-year-old physicist died of a disease contracted in the trenches.

The most remarkable stories waiting to be found inside Einstein's equations reveal themselves when we take an audacious and seemingly reckless leap. Instead of confining ourselves to describing the spacetime around spherical blobs of matter, why not think a little bigger? Why not try to use Einstein's equations to tell us about all of spacetime? Why can't we apply General Relativity to the entire universe? Einstein noticed this as a possibility very early in the development of his theory, and in 1917 he published a paper entitled 'Cosmological Considerations of the General Theory of Relativity'. It's a big step, of course, from thinking about someone falling off a roof to telling the story of the universe, and Einstein appears to have been uncharacteristically wobbly. In a letter to his friend Paul Ehrenfest a few days before he presented his paper to the Prussian Academy, he wrote 'I have ... again perpetrated something about gravitation theory which somewhat exposes me to the danger of being confined in a madhouse.'

The universe modelled in Einstein's 1917 paper is not the one we inhabit, but the paper is of interest for the introduction of what Einstein later came to view as a mistake. Einstein tried to find a solution to his equations that would describe a finite universe, populated by a uniform distribution of matter, and stable against gravitational collapse. At the time, this was a reasonable thing to do, because astronomers knew of only a single galaxy – the Milky Way – and the stars did not appear to be collapsing inwards towards each other. Einstein also seems to have had a particular story in mind; he felt that an eternal universe was more elegant than one that had a beginning, which left open the thorny question of a creator. He discovered, however, that General Relativity does not allow for a universe with stars, planets and galaxies to be eternal. Instead, his solution told the story of an unstable universe that would collapse inwards. Einstein tried to solve this unfortunate problem by adding a new term in his equations known as the cosmological constant. This extra term can act as a repulsive force, which Einstein adjusted to resist the tendency of his model universe to collapse under its own gravity. Later, he is famously said to have remarked to his friend George Gamow that the cosmological constant was his biggest blunder.

As physicists began to search for solutions to Einstein's equations, more and more possible universes were discovered. None, with the exception of Einstein's universe and a universe without matter and dominated by a (positive) cosmological constant discovered in 1917 by Willem de Sitter, was static. We will return to de Sitter's universe in a moment, but in every other case, Einstein's equations seemed to imply continual evolution, whereas Einstein himself felt that the universe should be unchanging and eternal. As more physicists worked with the equations, things only got worse for Einstein's static, eternal universe.

The first exact cosmological solution of Einstein's equations for a realistic universe filled with galaxies was discovered by Russian physicist Alexander Friedmann in 1922. He reached his result by assuming something that takes us all the way back to the beginning of this chapter; a Copernican universe in the sense that nowhere in space is special. This is known as the assumption of homogeneity and isotropy, and it corresponds to solving Einstein's equations with a completely uniform matter distribution. This may seem to be a gross oversimplification, and in the early 1920s the extent to which this assumption agreed with the observational evidence – a universe seemingly containing just a single galaxy – was tenuous. From a theoretical perspective, however, Friedmann's assumption makes perfect sense. It's the simplest assumption one can make, and it makes it relatively easy to do the sums! So relatively easy, in fact, that Friedmann's work was replicated and extended quite independently by a Belgian mathematician and priest named Georges Lemaître. Lemaître planted his flag firmly in the no-man's-land between religion and science – a strip of intellectual land occupied, whether we like it or not, by cosmology. A student of Harlow Shapley, this deeply religious man never saw a conflict between these two very different modes of human thought. He embodies the much debated and criticised modern notion, introduced by the evolutionary biologist Stephen J. Gould, that science and religion are non-overlapping magesteria, asking the same questions but operating within separate domains. My view is that this is far too simplistic a position to take; questions concerning the origin of the physical universe are of the same character as questions about the nature of the gravitational force or the behaviour of sub-atomic particles, and answers will surely be found by employing the methodology of science. Having said that, I am willing to recognise that romance, or wonder, or whatever the term is for that deep feeling of awe when contemplating the universe in all its immensity, is a central component of both religious and scientific experience, and perhaps there is room for both in providing the inspiration for the exploration of nature.

At least this is what Lemaître felt, and he used his twin perspectives as a guide on his intellectual journey through the cosmos throughout his distinguished career. Ordained a priest in 1923 while studying at the Catholic University in Louvain, Lemaître studied physics and mathematics alongside some of the great physicists and astronomers of the time, including Arthur Eddington and Harlow Shapley, from the University of Cambridge to Harvard and MIT, before returning to Belgium in 1925 to work with Einstein's general relativity.

Lemaître never met Alexander Friedmann, who died from typhoid in 1925. They never spoke or corresponded, and Lemaître was almost certainly unaware of the obscure paper Friedmann had published describing a dynamic and changing universe. He followed the same intellectual path, however, assuming an isotropic and homogeneous distribution of matter in the cosmos, and searching for solutions to Einstein's equations that describe the story of this smooth and uniform universe. And, of course, he came to the same conclusion; such a universe cannot be static – it must either expand or contract. Lemaître met

Einstein at the 1927 Solvay Conference in Brussels, and told him of his conclusions. 'Your calculations are correct, but your physics insight is abominable', snapped the great man. Einstein was wrong. By 1931, Lemaître was writing papers containing wonderfully vivid phrases and making clear his view that Einstein's theory requires a moment of creation – a Big Bang. He writes of 'a day without yesterday', and of the universe emerging from a 'primeval atom'.

In 1934, the Princeton physicist Howard Percy Robertson catalogued all of the possible solutions to Einstein's equations given a uniform distribution of matter throughout the cosmos – a perfect Copernican principle according to which no place in the cosmos is special or significant. The models containing matter tend to describe either an expanding or contracting universe, and therefore suggest a quite wonderful thing; there may have been a day without a yesterday. Einstein's equations contain within them a scientific creation story, even though their author himself resisted it.

The story of Einstein's Theory of General Relativity, and its subsequent application to the whole universe, delivers a compelling narrative illustrating the power of physics. The theory, inspired by thinking about a man falling off a roof, predicts that there was a moment of creation. No experimental measurements are required and no observations need be made other than that things fall at the same rate in a gravitational field. There are multiple layers of irony here! The idea that such progress towards answering the most profound questions

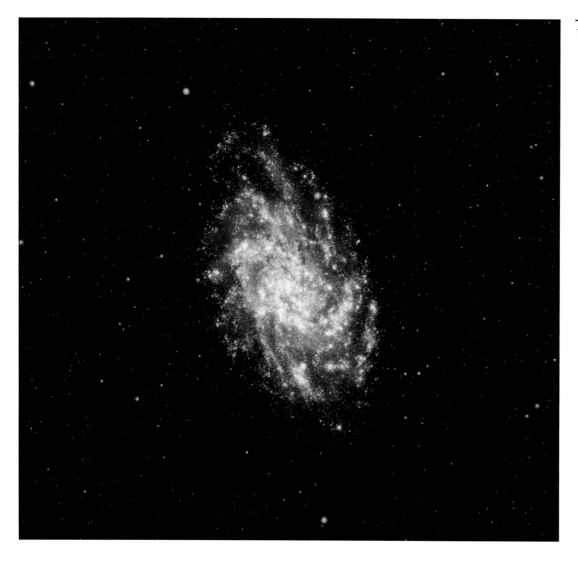

TRIANGULUM GALAXY (M33)
Also known as the Pinwheel galaxy, the Triangulum galaxy is a spiral galaxy which lies about 2.4 million light years from Earth.

about our origins can be made by thinking alone is almost Aristotelian: a partial throwback to the lofty authority of the classical world that Bruno, Copernicus and Galileo did so much to overturn. That the equations seem to describe a universe with a necessary moment of creation, lending support, at least in Lemaître's eyes, to the notion of a creator, would also appear to bring us full circle and back to Borman, Lovell and Anders and the creation stories of old. Indeed, Pope Pius XII, on hearing about the new cosmology, said 'True science to an ever increasing degree discovers God, as though God was waiting behind each door opened by science'. Einstein, to his deep chagrin, having thrown a blanket of rational thought across a landscape of mythology, appeared to have replaced one creation story with another.

To finish the story of our magnificent relegation, let me briefly address these points. The theoretical prediction of an expanding universe does of course require experimental verification, and this came rapidly. On 15 March 1929, Edwin Hubble published a paper entitled 'A relation between distance and radial velocity among extra-galactic nebulae', in which he reported his observation that all galaxies beyond our local group are rushing away from us. Moreover, the more distant the galaxy, the higher its speed of recession. This is precisely what an expanding universe as predicted by Einstein's theory should look like. In 1948, Alpher, Bethe and Gamow published a famous paper (with the coolest author list in the history of physics) which showed how the observed abundance of light chemical elements in the universe could be calculated assuming a very

GALAXIES BEYOND OUR OWN
This Hubble Space Telescope image shows distant galaxies beyond our own which as yet are unidentified and unexplored by scientists on Earth. It is a challenge to our future generations to uncover their secrets.

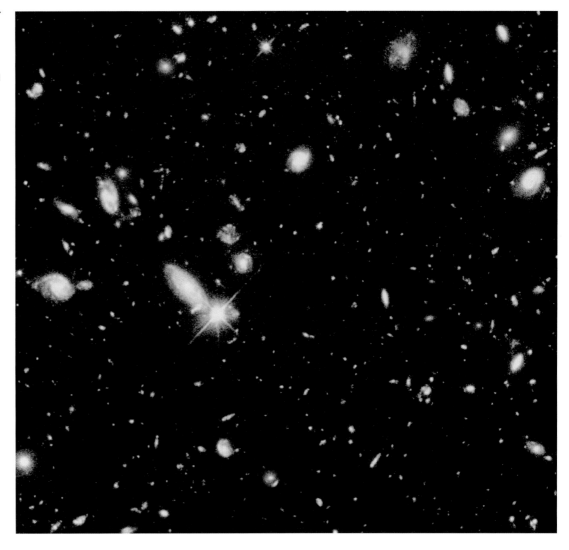

hot, dense phase in the early history of the universe. Modern calculations of these abundances are extremely precise, and agree perfectly with astronomical observations. Perhaps most compellingly of all, the afterglow of the Big Bang, known as the Cosmic Microwave Background Radiation, also predicted by Alpher and Herman in 1948, was discovered by Penzias and Wilson in 1964. We will have much to say about the Cosmic Microwave Background in the following chapters; for now, it is sufficient to say that the discovery that the universe is still glowing at a temperature of 2.7 degrees above absolute zero was the final evidence that convinced even the most sceptical scientists that the Big Bang theory was the most compelling model for the evolution of the universe.

What, though, of the thorny question of the cause of the Big Bang itself? What was the origin of Lemaître's primeval atom? Did God really do it? The standard Big Bang cosmology of the twentieth century has no answer to this question, but twenty-first-century cosmology does. We will address the current scientific understanding of what happened before the Big Bang later on, but let me offer a tantalising hint here. It is now thought that before the Big Bang the universe underwent a period of exponential expansion known as inflation. In this time, the universe behaved in accord with de Sitter's matter-less solution to Einstein's

OUR OLDEST LIGHTS
This snapshot of our universe
reveals the oldest lights
within it. The tiny temperature
fluctuations that ripple through
the skies reveal the presence of
the stars and galaxies of today
and for the future.

equations discovered in 1917. This period of rapid expansion gave us the homogeneous and isotropic distribution of matter we see today on large distance scales, which is the reason why Friedmann and Lemaître's simple Copernican assumptions lead to a description of the evolution of the universe after the Big Bang that fits observational data perfectly. There are no special places in the universe because the early inflationary expansion smoothed everything out. When inflation stopped, the energy contained within the field that drove it was dumped back into the universe, creating all the matter and radiation we observe today. Small fluctuations in the inflation field seeded the formation of the galaxies, uniformly distributed across the sky in their billions, each containing countless worlds, quite possibly without end beyond the visible horizon. In the words of Georges Lemaître, 'Standing on a well-cooled cinder we see the slow fading of the suns and we try to recall the vanished brilliance of the origin of the worlds.' Our cinder is not special; it is insignificant in size; one world amongst billions in one galaxy amongst trillions. But it has been a tremendous ascent into insignificance because, by the virtuous combination of observation and thought, we have been able to discover our place. How Giordano Bruno would have loved what we found.

OUR OLDEST LIGHTS
This snapshot of our universe reveals the oldest lights within it. The tiny temperature fluctuations that ripple through the skies reveal the presence of the stars and galaxies of today and for the future.

ARE
WE
ALONE?

Sometimes I think we are alone in the universe
and sometimes I think we're not.
In either case the idea is quite staggering.

Arthur C. Clarke

SCIENCE FACT OR FICTION?

There are questions to which knowing the answers would have a profound cultural effect. The question of our solitude is one. Are we alone in the universe – yes or no? One of these is true. The question as posed isn't a good one, however, because it is impossible to answer in the affirmative. We have no chance, even in principle, of exploring the entire universe, which extends way beyond the visible horizon 46 billion light years away. The answer can therefore never be yes with certainty. Indeed, if the universe is infinite in extent, we have our answer! No, we are not alone. The laws of nature self-evidently allow life to exist, and no matter how improbable, life must have arisen an infinite number of times. In itself, this is quite a challenging statement, and we will explore it in more detail in later on. But this isn't really what most of us want to know.

I've always been interested in aliens – the ones that fly spaceships around – and I want to talk to one. On a winter afternoon in 1977 I stood in a queue that went around three sides of the Odeon cinema in Oldham with my dad, shuffling through half-frozen puddles to see *Star Wars*, and spent the next decade building Millennium Falcons out of Lego. At some point in 1979 I picked up a magazine about *Alien*, and moved on to *Nostromo*, which required more bricks. To my delight I saw *Alien* when I was 11 years old at Friday Evening Film Society at school, and it didn't put me off. I just realised I really liked the spaceships, and didn't care much about the organic stuff. Everyone should see *Alien* at 11. To hell with the ratings; terror, technology and Sigourney Weaver are good for the soul.

Science fiction was a natural home for my imagination. I'd been interested in astronomy for a while, I'm not sure why, but the study of the stars seemed clean and precise and romantic; something done on cold nights before Christmas with mittens and imagination. *Star Wars, Star Trek, Alien*, Arthur C. Clarke and Isaac Asimov were merged seamlessly with Patrick Moore, Carl Sagan and James Burke, and they remain so; fact and fiction are inseparable in dreams. The superficially orthogonal desires to do science and to imagine distant worlds are closely related: shadows cast by different lights.

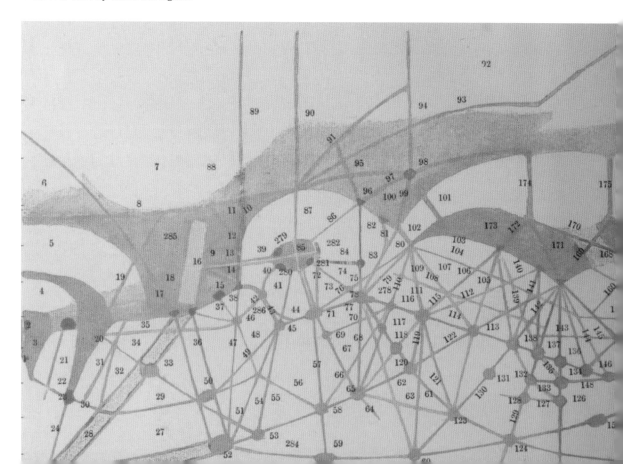

So the question 'Are we alone in the universe?' might make good science fiction, but it is not well posed in a scientific sense because the universe is too big for us to explore in its entirety. If we restrict the domain of the question, however, we can address it scientifically. 'Are we alone in the solar system?' is a question we are actively seeking to answer with Mars rovers and future missions to the moons of Jupiter and Saturn, where the conditions necessary for life may be present on multiple worlds. But even here, the use of the word 'alone' in the question is problematic. Would we be alone if the universe were full of microbes? Would you feel alone stranded in a deep cave with no means of escape and a billion bacteria for company? If not being alone means having intelligent beings to communicate with – sophisticated creatures that build civilisations, have feelings, do science and respond emotionally to the universe, then we have our answer in the solar system. Yes – Earth is the only world that is home to a civilisation, and we are alone.

How far might we reasonably expect to extend the domain of our question beyond the solar system? I find it impossible to believe that we'll ever explore the universe beyond our own galaxy. The distance between the Milky Way and our nearest neighbour, Andromeda, is over 2 million light years, and that seems to me to be an unbridgeable distance, at least given the known laws of physics. But that still leaves an island of several hundred billions of stars, 100,000 light years across. We will therefore rephrase our question so that we have a chance of interrogating it in a scientific way, and ask 'Are we the only intelligent civilisation in the Milky Way galaxy?' If the answer is yes, then we are in the cosmic equivalent of an inescapable cave and that would have made my 11-year-old self, gazing up at a dark sky of infinite possibilities, extremely sad. There may be others out there amongst the distant galaxies, but we'll never know. If the answer is 'No', on the other hand, this would have profound consequences. Aliens would exist in a truly science-fiction sense; beings with spacecraft, culture, religion, art, beliefs, hopes and dreams, out there amongst the stars, waiting for us to speak with them. What are the chances of that? We don't know, but at least we have posed a question that can be explored scientifically. How many intelligent civilisations are there likely to be in the Milky Way, given the available evidence today?

CANALS ON MARS?
American amateur astronomer Percival Lowell's map of Mars shows what became popularly known at the time as canals on the planet's surface. Lowell believed these features were created by intelligent life forms.

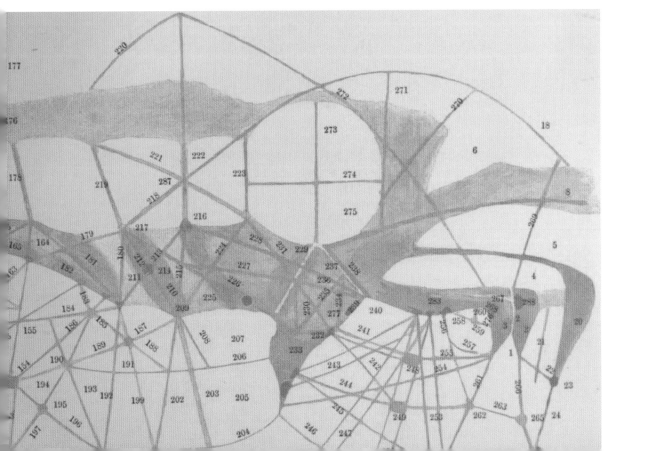

THE FIRST ALIENS

On 24 June 1947, Ken Arnold, an amateur pilot from Scobey, Montana, was flying over Mount Rainier, one of the most dangerous volcanoes in the world. Arnold was an experienced pilot with thousands of flying hours, and this implied he was a trustworthy observer. On returning to the airfield, he claimed to have seen nine objects flying in the mountain skies, describing them as 'flat like a pie pan' and 'like a big flat disc'. He estimated the discs were flying in formation at speeds of up to 1920 kilometres per hour. The press jumped on the story – coining the term 'flying saucer' – and within weeks hundreds of similar sightings were reported from all over the world. On 4 July a United Airlines crew reported seeing another formation of nine discs over the skies of Idaho, and four days later, the mother of all UFO stories exploded at Roswell, New Mexico, with the confirmation and then rapid retraction by the United States Air Force of a recovered 'flying disc' – an alien craft crash-landed on Earth.

I'll put my cards on the table here; I believe in UFOs. That is to say, I believe that there have been sightings of flying things in the sky that the observers were unable to identify, some of which were objects. But I do not believe for a moment that these were spacecraft flown by aliens. Occam's razor is an important tool in science. It shouldn't be oversold; nature can be complex and bizarre. But as a rule of thumb, it is most sensible to adopt the simplest explanation for an observation until the evidence overwhelms it.

My favourite response to the criticism that dismissing the possibility of alien visitations to Earth is unscientific was provided by physicist and

THE FIRST FLYING SAUCERS
Ken Arnold's original letter about sightings of UFOs, as he sent it to Army Air Force Intelligence on 12 July 1947.

Nobel Laureate Richard Feynman in his Messenger Lectures at Cornell University in 1964: 'Some years ago I had a conversation with a layman about flying saucers — because I am scientific I know all about flying saucers! I said "I don't think there are flying saucers". So my antagonist said, "Is it impossible that there are flying saucers? Can you prove that it's impossible?" "No", I said, "I can't prove it's impossible. It's just very unlikely". At that he said, "You are very unscientific. If you can't prove it impossible then how can you say that it's unlikely?" But that is the way that is scientific. It is scientific only to say what is more likely and what less likely, and not to be proving all the time the possible and impossible. To define what I mean, I might have said to him, "Listen, I mean that from my knowledge of the world that I see around me, I think that it is much more likely that the reports of flying saucers are the results of the known irrational characteristics of terrestrial intelligence than of the unknown rational efforts of extraterrestrial intelligence." It is just more likely. That is all.'

Irrespective of the veracity of the stories of mutilated cows, crop circles and violated Mid-Westerners at the hands of these alien visitors, the cultural impact of these early sightings was very real. America quickly entered into a media-fuelled love affair with alien invaders in shiny discs brandishing anal probes (why didn't they use MRI scanners, a non-Freudian would surely ask?). Of all the hundreds of thousands of references to flying saucers that began to appear in the media, a cartoon by Alan Dunn published in the *New Yorker* magazine on 20 May 1950 found its way into the lunchtime conversation of a group of scientists at the Los Alamos National Laboratory in New Mexico.

A LOAD OF RUBBISH?
Alan Dunn's cartoon in the *New Yorker* on 20 May 1950 blamed the aliens who had been increasingly 'seen' in New York for stealing city residents' rubbish bins.

Enrico Fermi was one of the greatest twentieth-century physicists. Italian by birth, he conducted his most acclaimed work in the United States, having left his native country with his Jewish wife Laura in 1938 as Mussolini's grip tightened. Fermi worked on the Manhattan Project throughout World War Two, first at Los Alamos, and then at the University of Chicago, where he was responsible for Chicago Pile 1, the world's first nuclear reactor. In a squash court underneath a disused sports stadium in December 1942, Fermi oversaw the first man-made nuclear chain reaction, paving the way for the Hiroshima and Nagasaki bombs.

After the war Fermi settled as a professor in Chicago, but he often visited Los Alamos. During one of these visits, in the summer of 1950, Fermi settled down for lunch with a group of colleagues including Edward Teller, the architect of the hydrogen bomb, and fellow Manhattan Project alumni Herbert York and Emil Konopinski. At some point, talk turned to the recent reports of UFO sightings and the *New Yorker* cartoon, stimulating Fermi to ask a simple question that turned a trivial conversation into a serious discussion: 'Where are they?'

FERMI'S PARADOX

The Fermi Paradox is the apparent contradiction between the high probability of extraterrestrial civilisations' existence and humanity's lack of contact with, or evidence for, such civilisations.

Fermi's question is a powerful and challenging one that deserves an answer. It has become known as the Fermi Paradox. There are hundreds of billions of star systems in the Milky Way galaxy. Our solar system is around 4.6 billion years old, but the galaxy is almost as old as the universe. If we assume life is relatively common, and on at least some of these planets intelligent civilisations arose, it follows that there should exist civilisations far in advance of our own somewhere in the galaxy. Why? Our civilisation has existed for around 10,000 years, and we've had access to modern technology for a few hundred. Our species, Homo sapiens, has existed for a quarter of a million years or so. This is a blink of an eye in comparison to the age of the Milky Way. So if we assume we are not the only civilisation in the galaxy, then at least a few others must have arisen billions of years ahead of us. But where are they? The distances are not so vast that we cannot imagine travelling between star systems in principle. It took us less than a single human lifetime to go from the Wright Brothers to the Moon. What might we imagine doing in the next hundred years? Or thousand years? Or ten thousand years? Or ten million years? Even with rocketry technology as currently imagined, we could colonise the entire galaxy on million-year timescales. The Fermi Paradox simply boils down to the question of why nobody has done this, given so many billions of worlds and so many billions of years. It is a very good question.

LISTEN VERY CAREFULLY

LISTENING TO THE NEIGHBOURS
The invention of the radio led many people to think that we would soon be communicating with our neighbours – Mars being considered the most likely planet to harbour intelligent life.

For three days in 1924, William F. Friedman had a very important job. As chief cryptographer to the US Army, Friedman was used to dealing with National Security responsibilities, but from 21–23 August he was asked to search for an unusual message. On these dates Mars and Earth came within 56 million kilometres of each other, the closest the two planets had been since 1845, and they would not be so close again until August 2003. This offered the best opportunity since the invention of radio to listen in on the neighbours.

To make the most of the planetary alignment, scientists at the United States Naval Observatory decided to conduct an ambitious experiment. Coordinated across the United States, they conducted a 'National Radio Silence Day', with every radio in the country quietened for five minutes on the hour, every hour, across a 36-hour period. With this unprecedented radio silence and a specially designed radio receiver mounted on an airship, the idea was to make the most of the Martian 'fly-by' and listen in for messages, intentional or otherwise, from the red planet.

Conspiracy theories notwithstanding, William F. Friedman didn't decipher the first message from an alien intelligence, and the American public soon tired of the disruption to their news bulletins, but the principle of the experiment was sound. The idea that we might listen in to aliens had first been proposed 30 years earlier by the physicist and engineer Nikola Tesla. Tesla suggested that a version of his wireless electrical transmission system could be used to contact beings from Mars, and subsequently presented evidence of first contact. He wasn't right, but in 1896, one year before the publication of *War of the Worlds*, it was certainly a plausible claim. Tesla wasn't alone; other luminaries of the time shared his optimism, including the pioneer of long-distance radio transmission, Guglielmo Marconi, who believed that listening to the neighbours would become a routine part of modern communications. By 1921 Marconi was publicly stating that he had intercepted wireless messages from Mars, and if only the codes could be deciphered, conversation would soon begin.

The failure of the National Radio Silence Day brought a temporary halt to the organised search for extraterrestrial signals, and the idea dropped out of scientific fashion until the post-war flying saucer boom. One of the first scientists to make the search for ET scientifically acceptable again was Philip Morrison, a contemporary and colleague of Fermi. It is not known whether they discussed the Fermi Paradox directly, but the idea of answering it certainly played on Morrison's mind throughout the 1950s. At the end of the decade Morrison published a famous and influential paper with another of Fermi's collaborators, Giuseppe Cocconi, laying out the principles of using radio telescopes to listen for signals. 'Searching for Interstellar Communications' was published in the prestigious journal *Nature*, and proposed a systematic search of the nearest star systems on a very specific radio frequency – the so-called 21cm hydrogen line.

Morrison and Cocconi chose the hydrogen line because it is a frequency that any technological civilisation interested in astronomy will be tuned in to. Hydrogen is the most abundant element in the universe, and hydrogen atoms emit radio waves at precisely this frequency. If we could see these wavelengths with our eyes, the sky would be aglow, and this is why astronomers tune their radio telescopes to the 21cm line to map the distribution of dust and gas in our galaxy and beyond. If a technological civilisation wants to be heard, then under the assumption that anyone with any sense does radio astronomy, the 21cm line would be the most obvious choice for a message.

Morrison and Cocconi's paper inspired the birth of one of the most widely debated and controversial astronomical projects of modern

21CM LINE

Hydrogen atoms consist of two particles – a single proton bound to a single electron. Protons and electrons have a property called spin, which for these particular particles (known as spin ½ Fermions, named after Enrico Fermi himself) can take only one of two values, often called spin 'up' and spin 'down'. There are therefore only two possible configurations of the spins in a hydrogen atom; the spins can be parallel to each other – both 'up' or both 'down', or anti-parallel – one 'up' and one 'down'. It turns out that the parallel case has slightly more energy than the anti-parallel case, and when the spin configuration flips from parallel to anti-parallel, this extra energy is carried away as a photon of light with a wavelength of 21cm.

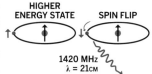

HIGHER ENERGY STATE   SPIN FLIP

1420 MHz
λ = 21CM

times. Within a year of its publication, the 85-foot radio telescope at the National Radio Astronomy Observatory in Green Bank, West Virginia, was pointing towards two nearby stars – Tau Ceti and Epsilon Eridani – listening in to the 21cm hydrogen line for any signs of unnatural order in the signals from the stars. The project, known as Ozma after a character from L. Frank Baum's *Land of Oz*, was the brainchild of Frank Drake, a young astronomer from Cornell University. Drake chose Tau Ceti and Epsilon Eridani as the first target star systems because of the stars' similarity to our own Sun and their proximity, just 10 and 12 light years away from Earth. In 1960 Drake had no idea if these stars harboured planetary systems, because no planets had been detected outside our solar system at that time. We now know that Drake's guess was a good one. Tau Ceti is thought to have five planets orbiting the star, with one of them in the habitable zone (see page 84). Epsilon Eridani is also thought to have at least one gas giant planet with an orbital period of around 7 years. After 150 hours of observation, Drake heard nothing, but for him this was the beginning of a lifetime dedicated to the search for extraterrestrial intelligence, a search commonly known by its acronym, SETI.

Today SETI is a global scientific effort, analysing data from telescopes used primarily for radio astronomy. The organisation also has a dedicated collection of telescopes designed specifically to detect signals from extraterrestrial civilisations at the Hat Creek Radio Observatory near San Francisco. The Allen Array, named after Microsoft founder Paul Allen who donated over $30 million to fund the construction of the project, consists of 42 radio antennae able to scan large areas of the sky at multiple radio frequencies, including the 21cm hydrogen line. If there are any civilisations making a serious attempt to contact us with technology at least as advanced as our own within a thousand light years, the Allen Array will hear them.

In the early 1960s, the scientific community was sceptical about such endeavours and Frank Drake was perceived as a maverick. It's important to be sceptical in science, but as Fermi understood, a back-of-the-envelope calculation with some plausible assumptions suggests that the search for ET may not be futile. Indeed, the alternative view that our civilisation is unique or extremely rare in a galaxy of a hundred billion suns appears outrageously solipsistic, and the sceptical finger might as easily be pointed at the cynics. There was, however, a handful of scientists who understood the importance of asking big questions, and together with Peter Pearman, a senior scientist at America's prestigious National Academy of Sciences, Drake organised the first SETI conference in November 1961. The Green Bank meeting was small, but the list of attendees, who named themselves The Order of the Dolphin, was impressive.

Philip Morrison was there, as was his co-author of the seminal 1959 *Nature* paper, Giuseppe Cocconi. I have a professional connection with Cocconi, who was a noted particle physicist and director of the Proton Synchrotron accelerator at CERN in Geneva. Cocconi was instrumental in discovering early experimental evidence for the pomeron, an object in particle physics known as a Regge trajectory that I have spent most of my career studying. The eminent, highly respected astronomer Otto Struve also attended. Struve publicly stated his belief in the existence of intelligent extraterrestrial life, perhaps because he had recently suggested a method for detecting alien planets outside our solar system (See page 85). Nobel Laureate Melvin Calvin, most famous for his work on photosynthesis, was present, along with future Hewlett Packard vice president for R&D Barney Oliver, astronomer Su-Shu Huang, communications specialist Dana Atchley and the colourful neuroscientist and dolphin researcher John Lilly. The most junior attendee was a 27-year-old postdoc. called Carl Sagan. I would love to have been there, although I'd have spent the whole time chatting with Cocconi about pomerons.

FIRST SETI CONFERENCE ATTENDEES

PETER PEARMAN
conference organiser

FRANK DRAKE

PHILIP MORRISON

DANA ATCHLEY
businessman and radio amateur

MELVIN CALVIN
chemist

SU-SHU HUANG
astronomer

JOHN C. LILLY
neuroscientist

BARNEY OLIVER
inventor

CARL SAGAN
astronomer

OTTO STRUVE
radio astronomer

GIUSEPPE COCCONI
particle physicist

TAU CETI

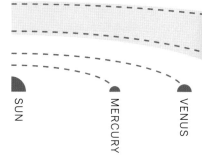

OUR SOLAR SYSTEM

SUN   MERCURY   VENUS

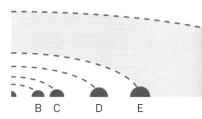

TAU CETI SYSTEM

B   C   D   E

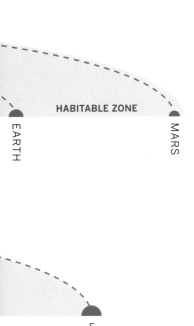

HABITABLE ZONE

EARTH

MARS

F

In preparation for the meeting, Drake drew up an agenda designed to stimulate a structured conversation amongst the group. If the search for intelligent extraterrestrial life was to be taken seriously, it was clear in Drake's mind that the discussion should be rigorous and provide a framework for future research. The way to do that is to address the problem quantitatively rather than qualitatively; to break it down into a series of probabilities that can be estimated, at least in principle, using observational data.

Drake focused on a well-defined question – the one we discussed above: how many intelligent civilisations exist in the Milky Way galaxy that we could in principle communicate with? Drake's brilliant insight was to express this in terms of a simple equation containing a series of probabilities. What is the fraction of stars in the galaxy that have planets? What is the average number of planets around a star that could support life? What is the fraction of those planets on which life begins? What is the probability that, given the emergence of simple life, intelligent life evolves? Given intelligence, how likely is it that the intelligent beings build radio telescopes and are therefore capable of communicating with us? Multiply all these probabilities together, and multiply by the number of stars in the Milky Way, and you get a number – the number of intelligent civilisations that have ever existed in the Milky Way.

This isn't all Drake did, however, because he was interested in the number of civilisations that we might be able to speak to now, and that requires the addition of a rather thought-provoking term – the average lifetime of civilisations from the moment they develop the technology to communicate. If a civilisation arose a billion years ago and vanished shortly afterwards, then we would never be able to talk to them. The question of the lifetime of a civilisation may have been more vivid in the early 1960s than it is today. The Manhattan Project had been the training ground for many of the great physicists, and the Cuban missile crisis was less than a year away, propelling the world, in Soviet Premier Khrushchev's words to President Kennedy, towards '... the abyss of a world nuclear-missile war'. To me, and to the participants at the Green Bank conference, the idea that a civilisation might destroy itself is both ludicrous and likely. We are pathetically inadequate at long-term planning, idiotically primitive in our destructive urges and pathologically incapable of simply getting along. More of this later! Putting the lifetime term into the equation was therefore scientifically valid and a political masterstroke; merely confronting the question should give us pause for thought at the very least.

To complete the equation with the lifetime term included – recall that it should give the number of currently contactable civilisations in the Milky Way – a little thought will convince you that the whole lot must be multiplied by the current rate of star formation in the galaxy. That might not be immediately obvious, but I have confidence you can demonstrate to yourself that it's the correct thing to do. Homework is good.

The completed equation, which is known as The Drake Equation, is shown to the right.

When Drake wrote down his equation, only R was known with precision. Star formation had been closely studied in parts of our galaxy and the data suggested a value of around one new star per year. The rest of the terms were unknown in the 1960s, and we will spend the majority of this chapter exploring them, given over 50 years of astronomical and biological research. Despite the lack of experimental data, however, the Green Bank participants spent the meeting debating each one of the terms in the Drake Equation. This is the power of Drake's formulation. It's not yet possible to make a measurement of the fraction of planets on which life emerges with any sort of precision, but it is possible to look at the experience we have on Earth, and increasingly in the wider solar

THE DRAKE EQUATION

$$N = R_* \times f_s \times f_p \times n_e \times f_l \times f_i \times f_c \times L$$

where:

N
the number of civilisations in our galaxy with which radio communication might be possible (i.e. which are on our current past light cone)

R_*
the average rate of star formation in our galaxy

f_p
the fraction of those stars that have planets

n_e
the average number of planets that can potentially support life per star that has planets

f_l
the fraction of planets that could support life that actually develop life at some point

f_i
the fraction of planets with life that actually go on to develop intelligent life (civilisations)

f_c
the fraction of civilisations that develop a technology that releases detectable signs of their existence into space

L
the length of time for which such civilisations release detectable signals into space

system, and make an informed guess. The probability of the emergence of intelligence given simple life is also a difficult question, but we do know that it took over 3 billion years on Earth, and that may give us a clue. Drake's equation is valuable therefore because it provides a framework for discussion and debate, focuses the mind and suggests a direction for future research, just as Drake intended.

The Green Bank meeting did produce a consensus number, based on the not inconsiderable expertise of the participants; there are of the order of 10,000 civilisations present now in the Milky Way with whom we could communicate if we had enough radio telescopes and the will to conduct a systematic search. Interestingly, Philip Morrison, veteran of the Manhattan Project, felt that the lifetime of technological civilisations may be so short that this number could well be zero, although he observed that '... if we never search, the chance of success is zero.'

I had the privilege of meeting Frank Drake during the filming of *Human Universe*. In my view he is one of the greatest living astronomers. Frank collects and cultivates orchids, and by complete coincidence I arrived at his house when his Stanhopea orchid was flowering. These delicate and complex flowers bloom for only two days every year, and the chance of seeing one on a random visit is therefore small. Frank turned to me and said 'well, so it is with SETI – we've learned that we must search over and over and over through the years, until we are in the right place at the right time to make the discovery'. There is 'hope' in its name, and there is nothing wrong at all with admitting a dash of hope.

Throughout the 1960s and 1970s, SETI projects both big and small continued to develop across the planet. Soviet scientists joined their American contemporaries in pointing radio receivers to the sky in the hope of detecting a signal in the noise. NASA considered funding Project Cyclops, a $10-billion-dollar super-array of 1500 dishes that could listen for signals originating up to 1000 light years from Earth. It never progressed beyond the planning stage, but the scale of the project demonstrates that SETI was considered to be a serious scientific endeavour. By the mid-1970s, various projects had come and gone but none had detected the faintest hint of a significant signal. This failure, combined with a lack of progress in pinning down any of the terms in the Drake Equation – it was not even certain that planets existed in large numbers beyond our solar system – made the search look increasingly futile. Not only was there a deafening silence, no one had much idea where to look or how hard to listen. NASA didn't lose faith, however, and in 1973 Ohio University's ten-year-old Big Ear telescope was optimised for a SETI survey and began taking data.

Four years later, on 18 August 1977, Jerry R. Ehman, then a volunteer at the Big Ear, received a knock on the door of his house. It was a Thursday morning and, as usual, standing at the door was a technician carrying reams of paper printouts. This was an age when a state-of-the-art hard disk could hold only a couple of megabytes, and every few days someone had to visit the telescope, print out the data and wipe the disks clean. Ehman put the three-days' worth of printed data onto his kitchen table and began searching. He was confronted with dozens of pages covered in hundreds of letters and numbers.

The list of numbers and letters depicts the strength of the signal hitting the telescope at different times. A space denotes low intensity, and higher intensities are registered as numbers from 0 to 9. For stronger signals still, letters between A and Z are used. Most of the data the 'Big Ear' recorded contained no letters; a stream of 1s and 2s signified sweeps across the general radio hiss of the sky. That morning, however, Ehman stumbled across something different. At approximately 10.16pm Eastern Standard Time on 15 August, a radio pulse of extreme intensity entered the antennae, recorded with the alphanumeric code 6EQUJ5. The signal lasted for

RADIO ASTRONOMY
The ambitious Square Kilometre Array (SKA) Project is part of a worldwide effort to develop the scientific goals and technical specifications for the next generation of radio observatories. In 2012, the SKA board announced that South Africa and Australia would share the location of the world's largest radio telescope; an array is seen here in a remote Northern Cape province of South Africa.

72 seconds, precisely the length of time a transmission of distant origin would register as the rotation of the Earth swept the telescope past the source. This is extremely important. If the signal had been caused by some kind of Earth-based interference, it would be highly unlikely to rise and fall in this manner, precisely and coincidently simulating the rotation of the Earth and the telescope's field of view on the sky. The peak was marked by the letter U, the strongest signal ever recorded by the Big Ear, denoting an intensity over 30 times that of the background emission of the galaxy. And equally strangely, the signal had a wavelength of 21cm – the hydrogen line favoured by Morrison and Cocconi in their 1959 *Nature* paper. A smoking gun for extraterrestrial communication?

With a now-famous flourish, Ehman circled the six characters and scribbled 'Wow!' on the printout. He then continued as a research scientist should, and looked to see if it happened again. He flicked through page after page, but the event of 10.16pm on 15 August was a solitary blip in the background noise. This presented a problem, because it should have happened again. The Big Ear telescope scans each part of the sky

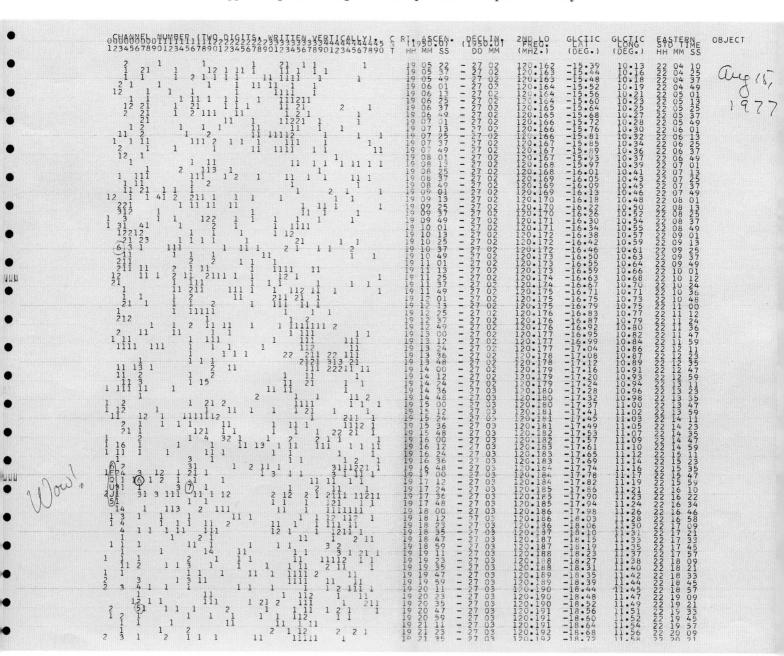

twice, separated by 3 minutes, so there should have been a similar Wow! signal in the data 3 minutes afterwards. None was present. This doesn't rule out an intelligent extraterrestrial origin; perhaps ET just turned the transmitter off a minute or so after it was first detected. Who knows?

The origin of the Wow! signal was narrowed down to a point in the sky in the direction of the constellation Sagittarius. Tau Sagittarii, a stable orange star twice the mass of our Sun and around 122 light years away, is the closest bright star to the source. Since August 1977 multiple attempts have been made to recover the signal using the world's most sensitive radio telescopes. Many hours have been spent listening, but nothing unusual has ever been detected again. Today, over 35 years later, there is no satisfactory explanation, but no serious scientist, no matter how embedded in SETI, would claim it as definitive evidence of intelligent extraterrestrial communication. Scientific results have to be repeatable, and the observation has never been repeated. For the moment, the Wow! signal remains an interesting anomaly in an otherwise silent sky. It is the stuff of dreams; the faintest of whispers in a great silence.

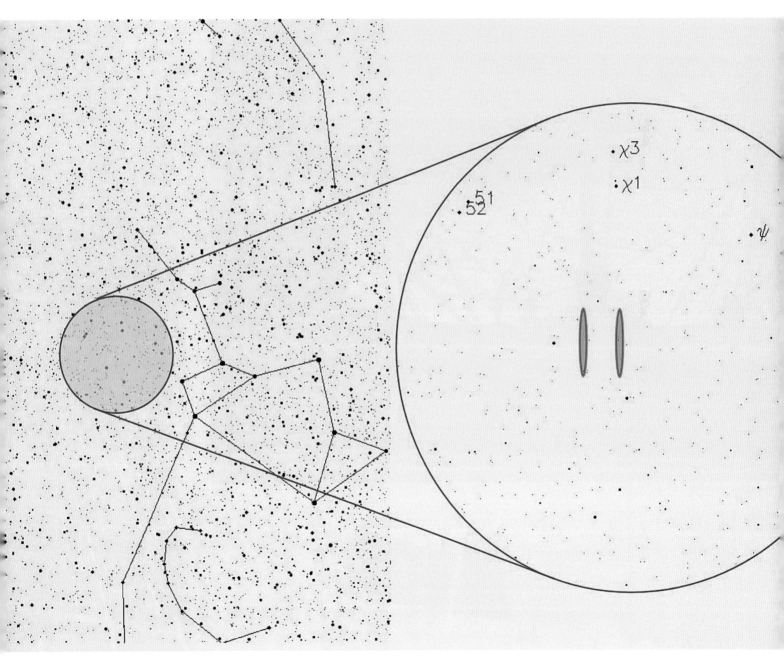

PROTOTYPE PROBE

This prototype of the two Voyager space probes launched in 1977 was created at NASA's Jet Propulsion Laboratory in Pasadena, California. In March 1977 the probe underwent a series of tests designed to see if the probe would survive launch. It passed all the tests and work began on the Voyager space probes which were to be sent out to explore the outer solar system gas giants, Jupiter, Saturn, Uranus and Neptune.

HIGH-GAIN ANTENNA

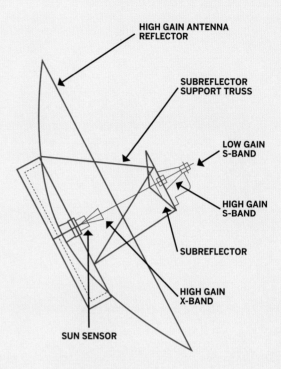

HIGH GAIN ANTENNA REFLECTOR

SUBREFLECTOR SUPPORT TRUSS

LOW GAIN S-BAND

HIGH GAIN S-BAND

SUBREFLECTOR

HIGH GAIN X-BAND

SUN SENSOR

PROBES ON TOUR

The Voyager probes have visited most of the outer planets on their way out of the solar system. Each visit has also used the planets' gravitational pull to slingshot the probes on their journey.

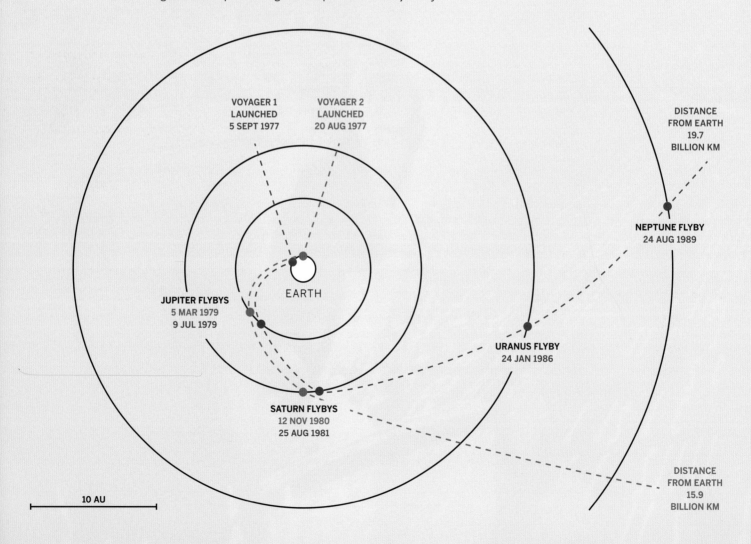

VOYAGER 1 LAUNCHED 5 SEPT 1977

VOYAGER 2 LAUNCHED 20 AUG 1977

EARTH

JUPITER FLYBYS
5 MAR 1979
9 JUL 1979

SATURN FLYBYS
12 NOV 1980
25 AUG 1981

URANUS FLYBY
24 JAN 1986

NEPTUNE FLYBY
24 AUG 1989

DISTANCE FROM EARTH 19.7 BILLION KM

DISTANCE FROM EARTH 15.9 BILLION KM

10 AU

THE GOLDEN VOYAGE

This is a present from a small, distant world, a token of our sounds,
our science, our images, our music, our thoughts and our feelings.
We are attempting to survive our time so we may live into yours.
US President Jimmy Carter

Two days after Jerry Ehman spotted the Wow! signal, the human race responded with a long-planned contribution to the interstellar conversation. In an explosive, serendipitous moment, the Voyager 2 spacecraft blasted into the sky above Space Launch Complex 41 at Kennedy Space Centre, followed two weeks later by its twin Voyager 1.

The Voyager missions were designed to take advantage of a rare planetary alignment to study the outer solar system gas giants Jupiter, Saturn, Uranus and Neptune. I remember the launch – I had collected a series of PG Tips tea cards called 'The Race Into Space', in which the Grand Tour mission was described as 'the most ambitious unmanned space project known'. Using the newly proposed gravity assist, a spacecraft could accelerate around Jupiter, Saturn and Uranus to encounter Neptune only a decade from launch. The Voyagers delivered, I suspect, way beyond their designers' wildest dreams, returning the first detailed pictures of the esoteric moons of Jupiter and Saturn, and in the case of Voyager 2, sweeping onwards to become the only spacecraft to date to visit Uranus and Neptune, where it photographed the distant ice moon Triton in the summer of 1989.

At the time of writing, on 8 July 2014, Voyager 1 is the most distant man-made object at over 127 astronomical units from Earth, so distant that radio waves take over 17½ hours to reach it. This puts Voyager 1 at the very edge of the solar system, on its way into interstellar space. The bus-sized spacecraft has enough electrical power to continue to communicate with its home world until around 2020, at which point it will fall silent. In 40,000 years it will drift within 1.6 light years of the red dwarf star Gliese 445 in the constellation of Camelopardalis. Voyager 2 will reach Sirius, the brightest star in the night sky, in 296,000 years.

The Voyagers are accompanied on their lonely flights out of our solar system by a dream – an unusually sentimental and hopeful afterthought to a scientific mission bolted to their sides almost 40 years ago.

The Voyager Golden Record is our message in a bottle. An old-fashioned phonograph record constructed of gold-plated copper floating through the universe, it contains what some would term a surreal mixture of sound recordings, images and information. It was designed to provide an alien civilisation with information about who we are, what we know and what our planet is like. There are 116 images on the disc; the first 30 or so are scientific, illustrating our solar system, our home world, the

WHAT THE DIAGRAM MEANT

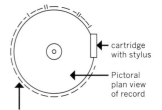

Binary code defining proper speed to turn the record (3.6 seconds) (| = Binary 1, — = Binary 0) Expressed in 0.70 x 10-9 seconds, the time period associated with the fundamental transition of the hydrogen atom

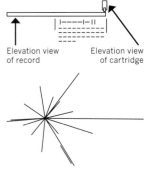

Elevation view of record   Elevation view of cartridge

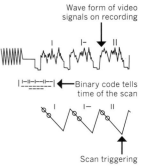

The location of our Sun utilising 14 pulsars of known directions from our Sun. Binary code defines frequency of pulses

VIDEO PORTION OF RECORDING

Wave form of video signals on recording

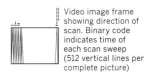

Binary code tells time of the scan

Scan triggering

Video image frame showing direction of scan. Binary code indicates time of each scan sweep (512 vertical lines per complete picture)

If properly decoded, the first image which will appear is a circle

Two lowest states of hydrogen atom. Vertical lines with dots indicate spin moments of proton and electron. Transition time from one state to the other provides the fundamental clock reference used in all the cover diagrams and decoded pictures

THE GOLDEN DISC

Our message to worlds beyond our own. The Voyager probes carried this phonograph on which were recorded sounds and images that would reflect life on Earth.

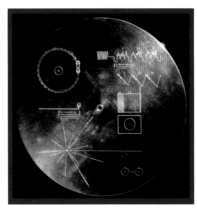

ARE WE ALONE?

structure of DNA, the anatomy of our bodies, our reproduction and our birth. Anatomy takes up more room than any other subject, perhaps reflecting our own fascination with what aliens might look like. In the most magnificently colloquial and futile gesture towards the alien's moral sensibilities, no nudity was allowed! I find it hard enough to imagine the inner workings of alien brains, but I cannot begin to fathom what it must be like inside the mind of a person who raised such an objection to the depiction of the human body. 'How do these beings reproduce? Perhaps they use those ten dangly things on the ends of their arms? Disgusting!'

The illustrations go on to detail our planet's landscapes and the variety of life on Earth, before dedicating 50 images to our lives and the civilisation we've constructed – from the Great Wall of China to a supermarket. Finally, there are images of the scientific instruments we have used to explore the universe from microscopes to telescopes, including the Titan rocket that launched the Voyagers into space. Chosen by a committee chaired by Carl Sagan, the disc also contains music and sounds, including human greetings in 55 languages, recordings reflecting 'the sounds of the Earth', and the ultimate 1977 mix tape featuring 90 minutes of music from Beethoven to Chuck Berry. Sagan wanted the Beatles' 'Here comes the Sun' on the disc, but EMI refused copyright permission for the universe. I like to imagine that Carl Sagan put the song on the record anyway in a great cosmic two-fingered salute to corporate Earth. That would have been pure Sagan – 'You're most welcome to go fetch it'.

The outside cover of the golden disc is more functional. As well as instructions on how to play back the images and sounds at precisely 16⅔ revolutions per minute for the audio, and how to build a record player, it also contains a map so that any extraterrestrial civilisation will be able to trace the record back to our planet. The map uses the position of 14 pulsars whose precise locations are marked relative to the Sun. The pulsars are identified by their fingerprints – each has a unique and unvarying rate of rotation. The most important piece of content on the cover is the key to unlock the information – a diagram illustrating the spin configurations of a hydrogen atom. The 21cm hydrogen emission line is a fundamental and universal property of nature, a Rosetta Stone that will allow an alien scientist to unlock the secrets of Earth. The disc also contains one last invisible source of information; electroplated onto the surface of the cover is an ultra-pure sample of uranium 238, an isotope with a half-life of 4.468 billion years. This is Voyager's clock, a way for any civilisation to determine the age of the record, assuming that they aren't creationists who disagree with radiometric dating. Perhaps these are the sorts of aliens that would also be offended by nudity.

For all the thought and care that went into these discs, neither Voyager spacecraft is heading towards any particular star; these tiny craft constructed by human hands will almost certainly never be found. The vastness of space swallows travellers, and of course Voyager's scientists and engineers knew this. That, however, is not the point; the act of launching these gilded emissaries into space expresses something important. It's my childhood science fiction dream of living in a *Star Wars* galaxy filled with life and possibilities. It is a desire to reach out to others, to attempt contact even when the chances are vanishingly small; a wish not to be alone. The golden discs are futile and yet filled with hope; the hope that we may one day know the boundaries of our loneliness and lay to rest the unsettling internal noise that accompanies the enduring silence.

Friends of space, how are you all? Have you eaten yet?
Come visit us if you have time.
Margaret Sook Ching, Voyager Golden Record

THE 1977 PLAYLIST

Brandenburg Concerto No. 2 in F
First Movement, Bach

'Kinds of Flowers'
Court gamelan, Java

Percussion
Senegal

Pygmy girls' initiation song
Zaire

'Morning Star' & 'Devil Bird'
Aborigine songs, Australia

'El Cascabel'
Mexico

'Johnny B. Goode'
Chuck Berry

Men's House Song
New Guinea

'Tsuru No Sugomori'
('Crane's Nest'), Shakuhachi, Japan

'Gavotte en rondeaux'
from the Partita No. 3, Bach

Queen of the Night aria, no. 14.
The Magic Flute, Mozart

'Tchakrulo'
Chorus, Georgian S.S.R.

Panpipes & Drum
Peru

'Melancholy Blues'
Louis Armstrong

Bagpipes
Azerbaijan S.S.R.

Rite of Spring
Stravinsky

The Well-Tempered Clavier
Book 2, Bach

Fifth Symphony
Beethoven

'Izlel je Delyo Hagdutin'
Bulgaria

Night Chant
Navajo Indians

'The Fairie Round'
Holborne, Paueans, Galliards, Almains and Other Short Aeirs

Panpipes
Solomon Islands

Wedding Song
Peru

'Flowing Streams'
Ch'in, China

'Jaat Kahan Ho'
Raga, India

'Dark Was the Night'
Blind Willie Johnson

String Quartet No. 13 in B flat
Beethoven

ALIEN WORLDS

An intrinsically improbable event may become highly probable if the number of events is very great.... [I]t is probable that a good many of the billions of planets in the Milky Way support intelligent forms of life. To me this conclusion is of great philosophical interest. I believe that science has reached the point where it is necessary to take into account the action of intelligent beings, in addition to the classical laws of physics.
Otto Struve

Let us now return to Frank Drake's equation and use it as intended as a framework to address in a systematic manner the question of our solitude. Recall that the equation consists of a series of terms which, when multiplied together, give an estimate of the number of currently contactable civilisations in the Milky Way galaxy. At the 1961 Green Bank meeting only the first term – the rate of star formation in the Milky Way – was known with any precision. Over half a century later, we can do much better. The next term in the equation is the fraction of stars in the Milky Way that have planets orbiting around them – most definitely a prerequisite for an intelligent civilisation to emerge. It's true that the civilisation may not have remained confined to its home world, and we will discuss this possibility later on. But it must be true that for life to emerge and evolve to the point where it can build spacecraft, a planet of some sort is required.

This space we declare to be infinite...
In it are an infinity of worlds of the same kind as our own.
Giordano Bruno, 1584

KEPLER-62

OUR SOLAR SYSTEM

SUN

MERCURY

VENUS

KEPLER-62 SYSTEM

B C D

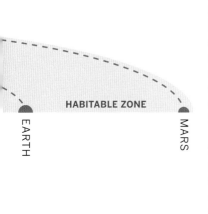

HABITABLE ZONE

EARTH

MARS

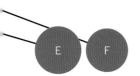

E   F

The existence of alien worlds has been speculated about for many centuries. Ever since Copernicus began the process of demoting our solar system from its preferred place in the cosmos, it has been natural to assume that at least some of the stars in the sky must have planetary systems. Yet despite this seemingly common-sense conclusion, reached by virtually every right-thinking astronomer from Giordano Bruno onwards, the existence of other planets remained nothing more than an educated guess well into my lifetime. The vast distances between the stars and the limitations of technology locked us inside our own solar system with no way of seeing beyond. Throughout the nineteenth century a number of astronomers claimed to have detected distant planets, but all these observations proved to be flawed.

Today the picture couldn't be more different; the night sky is known to be awash with worlds. One of the more enticing of the known solar systems is located around a slightly smaller, cooler version of our Sun called Kepler-62. About 1200 light years from Earth in the constellation of Lyra, the system has been widely studied because it has at least five planets. Two of them, Kepler 62-e and Kepler 62-f, are particularly interesting because they are Earth-like in both size and distance from the star. Bathed in Kepler-62-shine, these worlds may, if they have the right atmospheric conditions, support oceans of liquid water on their surfaces. We will discuss the significance of this in the context of life later on.

The discovery of extra-solar planets has been possible due to the rapid development of precision astronomical instruments, both space-based and terrestrial, that allow us to see beyond the bright glare of stars to the worlds that lie in the shadows. Imagine looking at our solar system from the nearest star system to Earth, Alpha Centauri. The system is 4.37 light years away, and consists of two sun-like stars – one slightly more massive than the other – orbiting each other with a period of approximately 80 years. The red dwarf Proxima Centauri is probably a distant gravitationally bound component of the system, making it a loosely bound triple star. Looking back towards Earth from 40 trillion kilometres with the naked eye, our sun would look like any other solitary star. Detecting exoplanets is no easy task because planets are vanishingly small and faint, masked by the brightness of their parent stars, and directly imaging them remains a major technical challenge.

To step out of the glare has required the development of indirect methods of detection based on surprisingly sensitive technologies. On 21 April 1992 the first conclusive detection of an exoplanet was made by radio astronomers Aleksander Wolszczan and Dale Frail, working at the Arecibo Observatory in Puerto Rico. They were hunting for planets around a pulsar known as PSR 1257+12, located 1000 light years from Earth, using a delicate method of indirect observation known as pulsar timing. Pulsars are spinning neutron stars, some of the most exotic objects in the universe. PSR 1257+12 is 50 per cent more massive than our Sun, but has a radius of just over 10 kilometres. It is, in effect, a giant atomic nucleus, spinning on its axis every 0.006219 seconds – that's 9,650 rpm. As you may gather from this rather precise statement, it is possible to measure the spin-rates of pulsars with great precision by timing the interval between pulses of radio waves emitted from the stars like a lighthouse. Wolszczan and Frail reasoned that if a large enough planet was orbiting a pulsar, the gravitational tug should shift the arrival times of the radio pulses by enough to be detectable. And sure enough, they found two planets orbiting PSR 1257+12, and measured their masses and orbits. Planet A has a mass of 0.020 times the mass of Earth and orbits the star once every 25.262 days. Planet B is 4.3 times the mass of Earth, and orbits once every 66.5419 days. Subsequently, a third planet has been discovered, with a mass of 3.9 times that of Earth and orbiting every 98.2114 days. Pulsar astronomy is indeed a precision science.

KEPLER-62F WITH 62E AS MORNING STAR

An artist's impression of the smallest habitable planet. Like our solar system, Kepler-62 has two habitable zone worlds. Kepler-62f is approximately 40 per cent larger than Earth; Kepler-62e, 60 per cent larger than Earth, orbits the inner edge of the habitable zone.

THE HABITABLE ZONE

The most important requirement for the evolution of life as we know it is liquid water. This can only exist on the surface of a planet if that planet is far enough away from the star at the centre of its planetary system: too close and the surface is too hot, resulting in any water boiling off into space; too far away and the surface is too cold and the water will exist only as ice. The too hot/ too cold scenario is what is known as the Goldilocks Zone. The distance and width of the Goldilocks Zone also depend on the size and temperature of the central star – it is further away from large, hot stars and closer in systems with small, cold stars. Using the Hertzsprung-Russell diagram (see page 97) and the known size of the star allows the calculation of each system's Goldilocks Zone, thus allowing us to determine whether the observed planets are likely to have liquid water and are therefore candidates for the evolution of life.

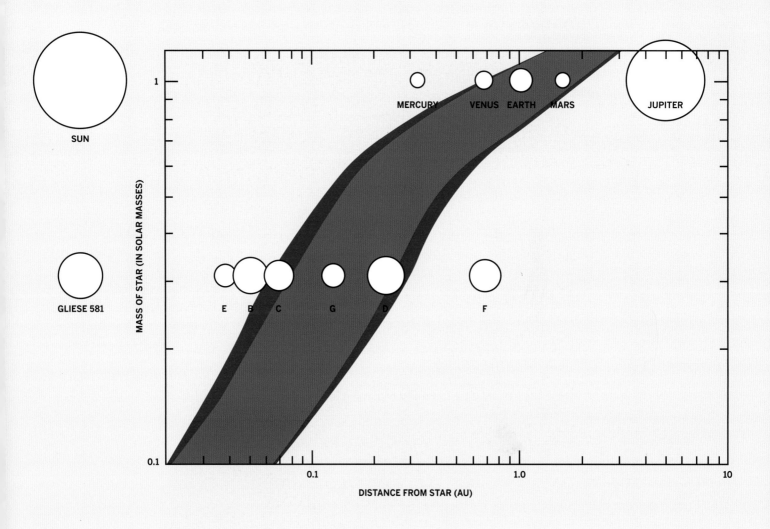

SUN

GLIESE 581

MASS OF STAR (IN SOLAR MASSES)

1

0.1

MERCURY   VENUS   EARTH   MARS   JUPITER

E   B   C   G   D   F

0.1   1.0   10

DISTANCE FROM STAR (AU)

HABITABLE ZONE   POSSIBLE EXTENSION OF ZONE DUE TO VARIOUS UNCERTAINTIES

This was an historic observation, but of limited direct interest to SETI since there is absolutely no chance that life could survive the hostile environment around such a violent astronomical object. It was, however, an existence proof – the first discovery of planets beyond our solar system, and a surprising one at that.

To search for Earth-like planets around Sun-like stars required the development of different but equally beautiful methods of observation. The first of these to be deployed was the radial velocity method. A star doesn't sit still at the centre of a solar system with planets orbiting around it. Rather, the star and planets orbit around their common centre of mass. The centre of mass of a solar system with a single star will always be inside the star itself, because it carries virtually all of the mass, but the star will still wobble around the centre of mass of the system as seen from Earth.

This planetary-induced wobble is small but measurable. In our solar system Jupiter causes our Sun to wobble backwards and forwards with a velocity change of approximately 12.4 m/s across a period of twelve years. The Earth's effect is minute in comparison, inducing a velocity change of just 0.1 m/s over a period of a year.

In the 1950s, future Green Bank pioneer Otto Struve suggested that such a planetary-induced wobble could be detected using the Doppler Effect. When a star moves towards the Earth, its light is shifted towards the blue part of the spectrum, and when it moves away from the Earth its light is shifted towards the red part of the spectrum. By making measurements of the specific frequencies (i.e. colours) of light absorbed by chemical elements in the star's atmosphere, and measuring how much

THE DOPPLER METHOD
Otto Struve (1897–1963) was a member of an astronomical dynasty. His father, uncle, grandfather and great-grandfather were renowned astronomers. He was awarded the Gold Medal of the Royal Astronomical Society in 1944, the fourth member of his family to be honoured in this way, for his work on stellar spectroscopy.

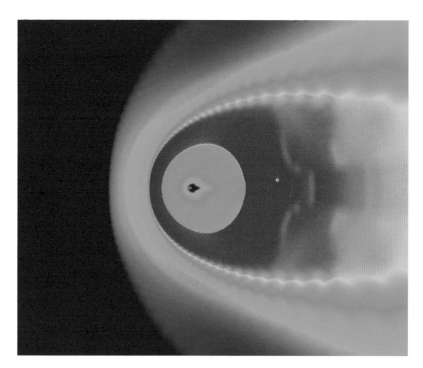

HELIOSPHERE MODEL
This Magnetohydrodynamical (MHD) model illustrates the interaction of the Sun's heliosphere (within the central blue dot) with the local interstellar medium (ISM), which traverses left to right.

these are shifted relative to the known frequencies as measured here on Earth, the motion of the star backwards and forwards can be determined over a period of time, and this can be used to calculate the orbital period of the planet and to estimate its mass. If there is more than one planet, the motion of the star will be more complicated, but since the orbital periods of the planets are regular, the contributions of the different planets to the star's wobble can be figured out.

Struve was one of the first respected scientists to publicly state his belief in extraterrestrial life. In the 1950s, however, the spectrographs used to measure red and blue shift were only able to detect velocity changes of a few thousand m/s, and at the Green Bank meeting he could only speculate that his technique would one day confirm his prejudice that planetary systems are common. Struve didn't live long enough to see his method applied, dying just two years after Green Bank, long before technology caught up with his ambition. It took until 1995 for two Swiss astronomers, Michel Mayor and Didier Queloz, to detect a planetary-induced Doppler shift using the Observatoire de Haute-Provence in France. The team discovered a planet orbiting the Sun-like star 51 Pegasi, located 50.9 light years from Earth.

This planet is named 51 Pegasi b, but its nickname is Bellerophon, after the mythological Greek hero who rode Pegasus, the winged stallion. Since its historic discovery, Bellerophon has been observed and examined in quite some detail, and it is no second Earth. It is a deeply hostile world, orbiting its parent star every four Earth days on a trajectory that takes it far closer than Mercury approaches our own Sun. Unlike Mercury, Bellerophon is a gas giant planet with a mass 150 times that of the Earth and a surface temperature approaching 1000 degrees Celsius. Although only half the mass of Jupiter, it may have a greater radius because the high surface temperature causes it to swell. Such exoplanets are known as Hot Jupiters – big enough and close enough to cause a significant wobble in their parent stars, which is why these types of worlds were discovered first by the early planet hunters.

The first evidence of a potential Earth-like planet arrived in 2007, when Stephan Audrey and his team at the European Southern Observatory in Chile announced the discovery of a planet around the red dwarf star Gliese 581, just over 20 light years from Earth. This was

RADIAL VELOCITY METHOD

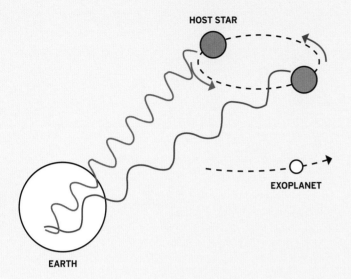

HOST STAR

EXOPLANET

EARTH

FINDING EXOPLANETS

One of the most exciting areas of current astronomical research is the hunt for planets around other stars – known simply as exoplanets – which are potential homes for extraterrestrial life. Until recently, such a search would have been impossible, as planets are too faint to see over interstellar distances. However, thanks to new instrumentation, we are now able to detect the telltale signals of exoplanets using two main techniques: the radial velocity method and the transit method. With these techniques, individual planets and even planetary systems have been discovered around hundreds of stars. Masses of these extrasolar planets range from a few times that of Earth to the size of 25 Jupiters.

TRANSIT METHOD

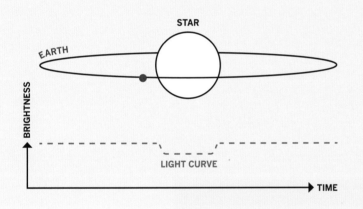

STAR

EARTH

BRIGHTNESS

LIGHT CURVE

TIME

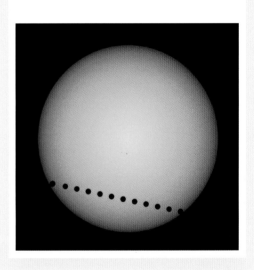

	KEPLER-4B	KEPLER-5B	KEPLER-6B	KEPLER-7B	KEPLER-8B
ORBITAL PERIOD (EARTH DAYS)	3.2 DAYS	3.5 DAYS	3.2 DAYS	4.9 DAYS	3.5 DAYS
EARTH MASS	4.31	18.8	15.0	16.9	18.3

FLUX: 1.000, 0.995, 0.990

PHASE (HOURS): -4, 0, 4

the second planet to be discovered in this system, but Gliese 581-c made headline news around the world because of its apparent Earth-like qualities. This planet is a rocky world, about five times as massive as Earth, and possibly the right distance away from its parent star to support liquid water on the surface: the stuff out of which science-fiction dreams are made. Further research has cast doubt on the idea that Gliese 581-b might have the necessary conditions to support life, but in March 2009 the second-Earth hunters got their own dedicated scientific instrument, and with it a cascade of new data became available.

The Kepler Space Telescope has transformed our knowledge of the distribution of planets in the Milky Way. Kepler is not a general-purpose instrument with multiple detectors and myriad ambitions; the telescope was designed for one purpose: to look for Earth-like planets. Free of the distorting effects of the Earth's atmosphere, Kepler carries a high-precision photometer, an instrument that has measured the light intensity from over 100,000 stars considered stable enough to support life on planets around them. Kepler searches for planets using a technique known as the transit method. If a planet passes across the face of a star as seen from Earth, the observed brightness of the star will drop by the tiniest of margins. Kepler's photometer is so sensitive it can measure changes in brightness (to use precise astronomical language we should say changes in the apparent magnitude) of less than 0.01 per cent. Observing repeated dips in brightness allows the orbital period of the planet to be measured, and the details of the changes in the brightness, combined with knowledge of the orbit, allows the size and mass of the planetary candidate to be estimated. The transit method has been extremely successful in the hunt for exoplanets, but the technique is not entirely reliable, often throwing up false positives. Once a promising candidate is found, the location is passed to ground-based telescopes for further analysis, and, if confirmed, the planets are classified as discoveries. Kepler has used the transit method of planet hunting on a quite extraordinary scale since it became fully operational in May 2009. As I write in July 2014, NASA's Exoplanet Archive lists 1,737 confirmed planets, over 50 per cent of which have been discovered using the Kepler data. This number is all the more staggering because Kepler is only capable of detecting a very small number of the planetary systems in our

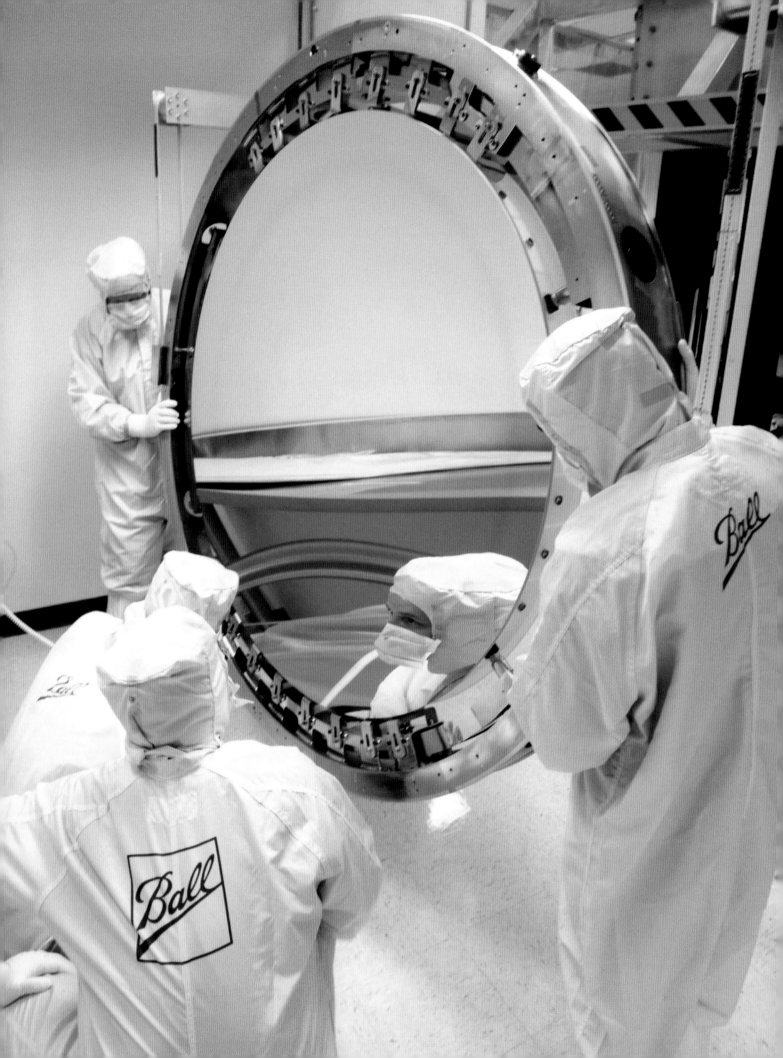

galaxy. Kepler views around 0.3 per cent of the sky in the constellations of Cygnus, Lyra and Draco, and even in this small patch, the telescope can only detect planets that pass directly in between their parent star and Earth. If the plane of the planetary orbits is orientated at the wrong angle, which is more likely than not, Kepler will not see any planets. Furthermore, Kepler only observed for four years, and because it has to see more than one transit to measure an orbit, it is blind to planets that orbit with periods greater than four years – which is the case for all the outer planets in our solar system. And finally, Kepler only sees stars out to a distance of approximately 3000 light years, whilst our galaxy has a diameter of 100,000 light years. Kepler's data set, then, contains only a tiny fraction of the planetary systems out there. All of these losses can be corrected for in a statistical sense, and when the numbers are crunched we have a reliable observation-based number to put into the Drake Equation. The fraction of stars that have planetary systems is close to 100 per cent! On average, there is at least 1 planet per star in the Milky Way galaxy, and we can insert the second term with confidence: $f_p = 1$.

The extraordinary Kepler mission was expected to last until 2016, but technical malfunctions may mean the telescope has now finished its planet-hunting activity. Even so, the huge volume of data is still being worked through and indications suggest it may have captured evidence for up to 3000 more planets circling distant stars.

This is encouraging for SETI enthusiasts, but in the hunt for civilisations, it's not the number of planets out there that really matters; rather, it is how many of these planets are capable of supporting life. This is the next term in the Drake Equation – the average number of planets per star that has planets that can support life – n_e. This is sometimes referred to as the Goldilocks question; how many of those billions of planets are not too hot and not too cold, but just right to allow life to exist on their surface?

KEPLER SPACE TELESCOPE FIELD OF VIEW
This star map shows the constellations of Cygnus and Lyra. The area marked with boxes indicates the telescope's field of view, which includes over 100,000 stars.

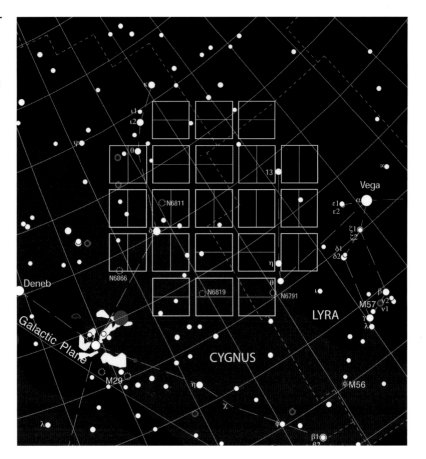

THE RECIPE FOR LIFE

*Extended regions of liquid water, conditions favourable for
the assembly of complex organic molecules, and energy sources
to sustain metabolism.*
NASA, 2008

Why Earth? What is it about our planet that makes it a home for life? In 2008 NASA brought together a team of scientists to define in the most basic terms the properties a planet needs to have a chance of supporting life, given our current scientific knowledge. Top of the list was liquid water – an ingredient virtually every biologist would agree is necessary for life. Water is a uniquely complex liquid, with its simple H_2O molecules forming great complexes held loosely together by hydrogen bonds. It forms the scaffolding around which biology happens, holding molecules and orientating them in just the right way for chemical reactions to take place. It is a superb solvent, and remains a liquid over an unusually large range of temperatures and pressures. It has been said that we will never truly understand biology until we understand water, such is its role in the chemistry of life on Earth. Fortunately, water is abundant in the universe. Hydrogen is the most common element, making up 74 per cent of the matter in the universe by mass. Oxygen is the third most abundant, at around 1 per cent, and these two reactive atoms combine to form water whenever they can. Water has been present in the universe for over 12 billion years, which we know because we've seen it. In July 2011, a giant reservoir of water was detected around an active galaxy known as APM 08279+5255. The cloud contains over 140 trillion times the amount of water in Earth's oceans, and is over 12 billion light years away, having formed less than 2 billion years after the Big Bang. So water is necessary for biology and, fortunately, extremely common throughout the universe.

Earth is unique in the solar system, however, because it is currently the only place where the surface conditions are right for water to exist in all three of its states: solid, liquid and gas. There are ice sheets at the poles and on the summits of the highest mountain peaks. In the atmosphere, clouds of water vapour form and fall as rain and snow, flowing back through rivers into the oceans that cover over 70 per cent of the surface. Mars has water, but on the cold red planet it can only be found as ice trapped in the poles and deep below ground and, just possibly, as sub-surface liquid lakes. Venus may once have been wet, but its proximity to the Sun and runaway greenhouse effect boiled any primordial oceans off into space long ago. This appears to suggest that it is Earth's distance from the Sun that defines its suitability for life. Drag the Earth closer to the Sun and the temperatures would rise, the oceans would evaporate into the atmosphere, and if things got too hot the water molecules would escape into space, leaving Earth a dry, Venusian world. Drag the Earth further out towards Mars, and temperatures would drop until eventually the surface water would freeze.

It might appear tempting, therefore, to look for planets at roughly the same distance from their stars as Earth in the search for living worlds. This would be oversimplistic, because things are a lot more complicated. The conditions on the surface of a planet depend on many factors, the distance to the star being only one. The mass of the planet determines the gravitational pull it exerts on the molecules in its atmosphere, and this determines which atmospheric molecules it can hang on to at a given temperature. This is important because the atmosphere plays a critical role in setting the surface temperature of a planet. Venus has the hottest surface in the solar system other than the Sun because of its greenhouse gas-laden atmosphere, despite being much further away from the Sun

THE HOTTEST STEP
A natural sinkhole, cut out of the limestone of the High Andes. Around 500–600 years ago, the Inca modified it by creating the circular steps. It is thought that it was a rudimentary agricultural research station, with every step having a range of microclimates, each of which was a habitable zone for their crops. An ideal TV location.

HYDROLOGICAL CYCLE
Ancient salt mines in an Incan Valley. The water comes from melting snow, which then evaporates leaving the salt. An example of water in its three states –solid ice, liquid water and water vapour.

than Mercury. The Moon, on the other hand, has very little atmosphere due to its small mass, and even though it is the same distance from the Sun as the Earth, its surface temperatures range from over 120°C in direct sunlight to below -150°C at night. NASA's Lunar Reconnaissance Orbiter measured the coldest temperature ever recorded in the solar system, -247°C, in the limb of a crater at the Moon's North Pole, which never receives sunlight because the Moon's spin axis is almost perpendicular to its orbital plane. The composition of the atmosphere is determined in part by the geology of the planet; on Earth, plate tectonics play an important role in regulating the amount of carbon dioxide in the atmosphere. CO_2 is a greenhouse gas, and higher concentrations of such gases raise the temperatures. The presence of sulphur dioxide in the atmosphere from volcanic eruptions can cool the surface of a planet, however, because sulphate aerosols reflect sunlight back out into space. The Mount Pinatubo eruption in June 1991 cooled the Earth's surface by up to 1.3 degrees for the three years following the eruption. And we shouldn't forget that life itself alters the composition of planetary atmospheres quite radically. Earth's atmosphere today is a product of the action of living things; before photosynthesis evolved, there was very little free oxygen in the atmosphere, and plants play an important role in removing CO_2 and locking it up in biomass. The planet's mass, spin axis, orbit, geology and atmospheric composition all conspire in a complex way to set the average surface temperature and atmospheric pressure, which ultimately determine whether liquid water can exist on the surface. And if life gets going, its effects have to be folded in as well.

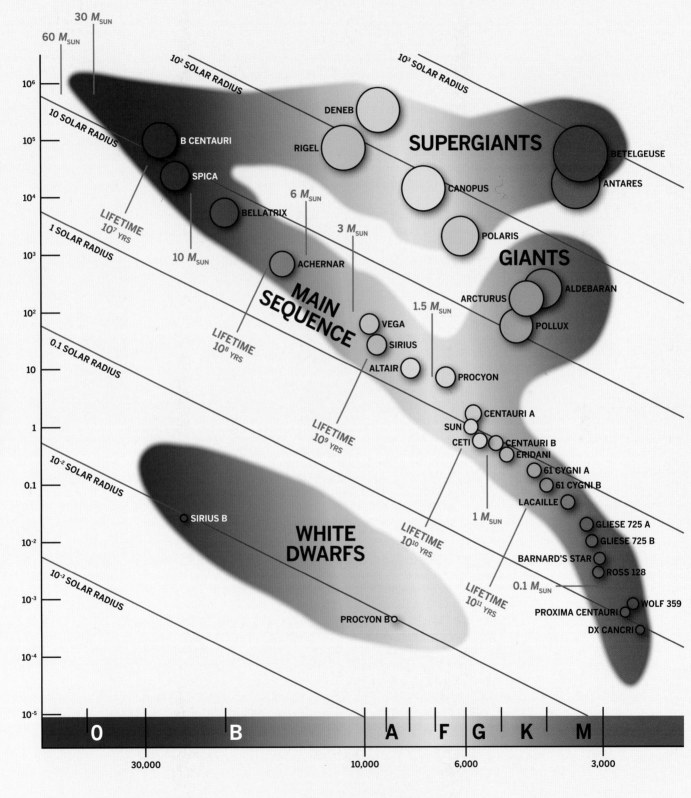

60 M_{SUN}
30 M_{SUN}

10^2 SOLAR RADIUS
10^3 SOLAR RADIUS

10 SOLAR RADIUS

10^6

DENEB

10^5
B CENTAURI
RIGEL

SUPERGIANTS

BETELGEUSE

SPICA
10^4
CANOPUS
ANTARES

LIFETIME
10^7 YRS

1 SOLAR RADIUS

6 M_{SUN}
BELLATRIX

3 M_{SUN}
POLARIS

10^3
10 M_{SUN}
ACHERNAR

GIANTS

MAIN
SEQUENCE

ARCTURUS
ALDEBARAN

10^2
1.5 M_{SUN}
POLLUX

LIFETIME
10^8 YRS

VEGA
SIRIUS

0.1 SOLAR RADIUS

10
ALTAIR
PROCYON

CENTAURI A

1
SUN
CETI
CENTAURI B
ERIDANI

10^{-2} SOLAR RADIUS

LIFETIME
10^9 YRS

61 CYGNI A
61 CYGNI B

0.1
LACAILLE

GLIESE 725 A
SIRIUS B
GLIESE 725 B

10^{-2}
BARNARD'S STAR
ROSS 128

WHITE
DWARFS
LIFETIME
10^{10} YRS
1 M_{SUN}
0.1 M_{SUN}

10^{-3} SOLAR RADIUS

10^{-3}
LIFETIME
10^{11} YRS
WOLF 359
PROXIMA CENTAURI

PROCYON B0
DX CANCRI

10^{-4}

LUMINOSITY (SOLAR UNITS)

10^{-5}

O
B
A
F
G
K
M

30,000
10,000
6,000
3,000

← INCREASING TEMPERATURE
SURFACE TEMPERATURE (KELVIN)
DECREASING TEMPERATURE →

Beyond the planet, a vitally important ingredient for producing a potentially living world is, of course, the parent star itself, and all stars are most definitely not alike. There are over two hundred billion stars in the Milky Way galaxy. The largest known supergiant stars are over 1500 times the diameter of our Sun. If such a star were located at the centre of our solar system, it would engulf Jupiter. At the other end of the spectrum are tiny red dwarfs, with diameters from around half that of our sun to as small as a tenth of it. The smallest known star at the time of writing goes by the name of 2MASS J05233822-1403022, which shines eight thousand times less brightly than our sun and is smaller (but denser) than Jupiter.

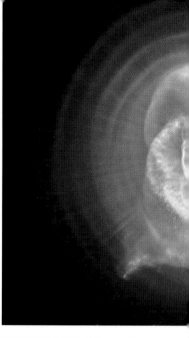

As with virtually everything in physics, a good way to make sense of this stellar menagerie is to draw a graph. The most famous graph in all of astronomy is known as the Hertzsprung-Russell diagram (on the previous page), after astronomers Ejnar Hertzsprung and Henry Norris Russell, who drew it independently in 1911. They plotted the surface temperature of the stars (which is directly related to their colour – hot stars are blue or white hot, cool stars are red) against their brightness. It is immediately obvious that the stars are not distributed randomly on the diagram. Most lie on a sweeping line ascending from the bottom right to the top left. This line is known as the Main Sequence. Our yellow sun lies around the middle of the main sequence, and all the stars on this line are generating their energy in the same way – by fusing hydrogen into helium in their cores. These are the 'standard stars', if you like, although their masses, lifetimes and suitability for the support of living solar systems are very different.

The basic physics underlying the Main Sequence line is simple. Stars are clouds of hydrogen and helium, which is pretty much all there is in the universe to a good approximation, collapsing under their own gravity. As the cloud collapses, it heats up. This is not surprising – all gases get hot when they are compressed – try pumping up a bicycle tyre. Eventually, the collapsing ball of gas gets so hot that the positively charged hydrogen atoms overcome their mutual electromagnetic repulsion and fuse together in a nuclear reaction to make helium. This releases a tremendous amount of energy, which further heats up the gas, increasing the rate of nuclear reactions and continuing to heat the gas. Hot gases want to expand, and so ultimately a balance will be reached between the crushing force of gravity and the outward pressure exerted by the nuclear-heated gas. This is the current state of our Sun, happily converting 600 million tonnes of hydrogen every second into helium to counteract the inward pull of gravity. For less massive stars, the equilibrium will be reached at a lower temperature because the inward pull of gravity is weaker. Having a lower surface temperature, these stars will be redder than our sun, and also less luminous. These are the dim, red stars at the bottom right of the diagram, known as red dwarfs. We've already met an example of a red dwarf – our nearest stellar neighbour, Proxima Centauri. Red dwarfs also have the longest lifetimes of the stars on the Main Sequence, simply because they have to burn their fuel at a lower rate in order to reach a stable equilibrium with gravity.

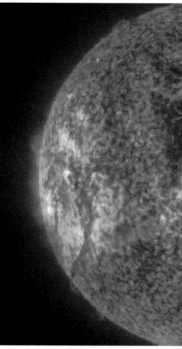

At the other end of the Main Sequence are the massive blue stars. Ten times the mass of our Sun or more, the inward pull of gravity is strong, and they have to burn their hydrogen fuel at a profligate rate to resist collapse. This makes them hot, and therefore blue, but also short-lived. The largest Main Sequence stars will use up their nuclear fuel in ten million years or less, at which point they will move off the Main Sequence to become red giant stars. The red giants, like the famous Betelgeuse in the constellation of Orion, are stars nearing the end of their lives. Starved of hydrogen in their cores, they begin to fuse helium into heavier elements like carbon and oxygen. These stars are the origin of

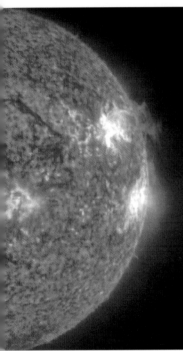

most of the heavy elements in your body. Their cores become superheated in their ultimately futile battle against gravity, causing their outer layers to expand and cool. This is why the red giants sit at the top right of the Hertzsprung-Russell diagram. They are vast, and therefore bright, but their cool surfaces cause them to glow a deep red. Red giants will last for only a few million years before they run out of nuclear fuel, at which point they shed their outer layers, forming one of the most beautiful sights in nature – a planetary nebula. It is these clouds, rich in carbon and oxygen, which ultimately distribute the building blocks of life into the galaxy. Your building blocks are likely to have been part of a planetary nebula at some point over five billion years ago. Cooling at the heart of the nebula is the fading core of the star, exposed as a white dwarf. These stars populate the bottom left of the Hertzsprung-Russell diagram.

There are a handful of other exotic stars out in the Milky Way. The vast blue supergiant stars like Deneb are extremely hot and extremely luminous. Deneb, the brightest star in Kepler's field of view in the constellation of Cygnus, is almost 200,000 times more luminous than our Sun, and 20 times more massive. It burns its nuclear fuel at a ferocious rate, and will probably explode in a supernova explosion within a few million years, leaving a black hole behind.

The Hertzsprung-Russell diagram, then, is the key to understanding stellar evolution, and also contains vital information for planet hunters. Stars that do not lie on the Main Sequence are highly unlikely to support planetary systems with the right conditions for life. They are either short-lived and ferociously bright, or have had a life history fraught with violence and change. The Main Sequence, containing the stable, hydrogen-burning stars, is where we should look for stability. But even there, the more massive, brighter stars are likely to be too short-lived for complex life to emerge. On Earth, life existed for over three billion years before complex organisms emerged in the Cambrian Explosion just 550,000 years ago. We will discuss the history of life on Earth in more detail a little later, but for now we might venture an educated guess that stars with lifetimes significantly shorter than a billion years or so are unlikely to preside over planets with intelligent civilisations. This rules out the blue stars at the top left of the Main Sequence. Even familiar stars like Sirius, the brightest star in the night sky and only twice the mass of the Sun, can probably be ruled out as its lifetime on the Main Sequence is expected to be a billion years at most. We are therefore left with stars on the Main Sequence with masses within a factor of two or less of our Sun as candidates for solar systems that could support complex life.

There may also be a lower limit on the masses of life-supporting stars, although this is very much an active area of research. Around 80 per cent of the stars in the Milky Way are red dwarfs, and many are known to have solar systems. Red dwarfs have potential lifetimes measured in the trillions of years, so there is no issue with their longevity. Despite their frugal use of fuel, however, red dwarfs tend to be volatile and variable in their light output. Sunspots can reduce their brightness by a factor of two for long periods of time, and violent flares can increase their brightness by a similar factor over time periods of days or even minutes. Planets in orbit around red dwarfs are therefore subject to significant and rapid changes in the amount of light and radiation they receive. Furthermore, because of their low light output, planets must be extremely close to the star if they are to be warm enough for liquid water to exist on the surface, irrespective of the details of their atmospheres. When planets orbit close to stars, they become tidally locked, with one hemisphere permanently facing the star and the other always facing into the darkness of space. We only see one face of our Moon for the same reason – tidal locking is inevitable for moons orbiting close to planets or planets orbiting close to stars. This results in a strange kind of climate for potentially habitable

THE CAT'S EYE NEBULA
Taken by NASA's Hubble Space Telescope, this image shows the sheer beauty of a dying star shedding its outer layers.

LIFE-GIVING STAR
The stable stars within the Main Sequence supply the heat and light that is required to sustain life on a planet such as Earth.

SUPERGIANT STAR
Antares – the white star in the bottom left of the picture – is one of the best-known examples of a supergiant star.

planets around red dwarf stars; there will be regions of permanent day, and regions of permanent night.

Despite all these problems, however, recent computer modelling suggests that red dwarf planets may be able to maintain stable surface conditions if they have thick, insulating atmospheres and deep oceans, and life has plenty of time to evolve in these unfamiliar (to us) conditions. The jury is still out as to whether the red dwarfs that populate the low-mass region of the Hertzsprung-Russell diagram could be candidates for living solar systems.

Where does all this leave us? If we take the conservative path, and focus our attentions on the Sun-like orange and yellow stars on the main sequence, we can look at the Kepler data to estimate how many of these so-called F, G and K-type stars in the Milky Way have rocky planets in the right orbits to allow liquid water to be present on the surface, at least in principle. These planets orbit within what is known as the habitable zone, and this is the number we want to measure and insert into the Drake Equation. This has been done, and the results are surprising. In a recent study, ten planets were identified as Earth-like in the Kepler data set, in the sense that they have the right mass and composition, and are in the right orbits around their parent Main Sequence F, G or K stars, to support liquid water on their surfaces for long periods of time. Applying all the statistical corrections to account for the alignment of the solar systems relative to Earth, the lack of ability to see planets with longer orbital periods, and so on, we can estimate with a reasonable degree of certainty that there are around 10,000 Earth-like planets capable of supporting life in Kepler's field of view. This in turn suggests that around a quarter of F, G and K stars in the Milky Way have potentially life-supporting planets in orbit around them, corresponding to ten billion habitable planets.

OUTSIDE THE GOLDILOCKS ZONE
Jupiter from Europa, where it is possible life exists in the sub-surface liquid water.

If we allow the possibility that planets around red dwarfs may also be habitable, then we can more than double that number.

There is one final point worth making about habitable zones around stars. In our solar system, Venus, Mars and Earth are within the habitable zone as commonly defined, but there are other places where life may exist. Several of the moons of Jupiter and Saturn are planet-sized worlds, and it is known that the Jovian satellites Europa and Ganymede, and quite possibly Saturn's giant moon Titan and the small but active Enceladus, have sub-surface oceans or lakes of liquid water. Europa in particular is considered to be one of the most likely places beyond Earth that may support life, even though it is outside the more commonly defined habitable zone around the Sun. If we admit the possibility that planet-sized moons may extend the habitable zone around stars, then the number of potentially life-sustaining worlds in the Milky Way increases significantly.

Over 50 years after the Green Bank meeting, the first three astronomical terms in the Drake Equation are now known from experimental data, and they are encouraging for SETI. There are, of course, large uncertainties, and one can find differing interpretations of the data in the academic literature. What is absolutely clear, however, is that the number of potential homes for life in the Milky Way is measured in hundreds of millions at the very least – most likely billions. From an astronomical perspective, the Milky Way could be teeming with life. The next three terms in the Drake Equation are biological; they concern the probability that life will emerge spontaneously on a planet that could support it, and the probability that the necessarily simple life that first appears evolves into complex, intelligent beings capable of constructing a technological civilisation. It is to these difficult questions that we now turn.

ARE WE ALONE?

ORIGINS

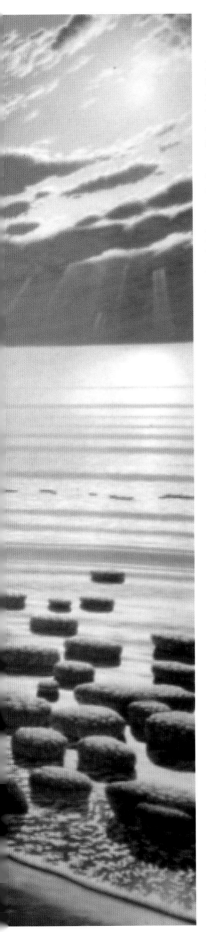

Earth formed 4.54 +/- 0.07 billion years ago out of the flattened disc of dust orbiting our young Sun. The planet was far from hospitable for the first few hundred million years of its life; it was an intensely hot and volcanic world, bombarded by asteroids and comets and, at least once, it collided with another planet, which resulted in the 23.5-degree tilt of our spin axis and the formation of the Moon.

Slowly, the solar system became a more ordered place, and Earth cooled to the point where liquid water could exist on its surface. There is evidence that liquid water existed as far back as 4.4 billion years, but it is certain that our planet was blue by the end of the late heavy bombardment 3.8 billion years ago, and around this time we find the first evidence of life. Structures known as microbially induced sedimentary structures were discovered in 2013 at a remote site in the Pilbara region of Western Australia. They were found in a sedimentary rock layer laid down in the early Archean period, 3.48 billion years ago. Similar structures are found today along ocean shorelines and in rivers and lakes, formed by the interaction of microbial mats with sediments carried through them by water currents. They indicate the presence of a complex microbial ecosystem, most likely a purple layer of slime that thrived in the warm, wet, oxygen-free environment of the early Earth, filling the atmosphere with the sulphurous stench of anaerobic breath. Early Earth would not appear welcoming to our eyes or noses.

Beyond 3.5 billion years, there is indirect evidence for the existence of life as far back as 3.7 billion years. Geologists studying some of the oldest sedimentary rocks on Earth in the Isua Supracrustal Belt in Western Greenland analysed the ratio of carbon isotopes in sedimentary rocks. The ratio of the heavier carbon 13 isotope to the more common carbon 12 can be used as a biomarker, because organisms preferentially use the lighter carbon 12 isotope in metabolic processes. Around 98.9 per cent of naturally occurring carbon is carbon 12, and if the concentration is significantly higher in a particular rock deposit then this is taken as evidence that the carbon was laid down by biological processes.

What can this evidence tell us about the probability of life emerging spontaneously on other worlds? The problem is that Earth is a sample size of one, so it would be erroneous to draw firm conclusions. It is interesting to observe that life emerged very early in the Earth's history – probably as soon as the conditions were right. The first half a billion years after Earth's formation is known as the Hadean Eon, named after the Greek god of the underworld. It is likely that the carbon dioxide atmosphere, volcanism and frequent bombardment from space made life impossible on the surface during the Hadean. From the start of the Archean Eon 4 billion years ago, and certainly after the violent period of the solar system's history known as the Late Heavy Bombardment – which is known from analysis of lunar rocks to have ended 3.8 billion years ago – Earth became a more stable planet, and this date coincides with the earliest evidence for life. It is tempting, therefore, to suggest that life began on Earth pretty much as soon as it could have done after the violence of its formation. If this is taken as a working hypothesis, then we might venture that the probability of life arising on a planet that could support it – the term f_l in the Drake Equation – is close to 100 per cent. This is, of course, speculative to say the least, and we would know this number with much greater certainty if we found that life arose independently on Mars, Europa, or one of the many bodies in the solar system that had or still have large bodies of liquid water on or below the surface. This is one of the most important motivations for the exploration of Mars and the moons of the outer solar system.

ARCHEAN LANDSCAPE
Earth as it would have appeared during the Archean period, 3.48 billion years ago.

A BRIEF HISTORY OF LIFE ON EARTH

At this stage in the analysis of the Drake Equation, it's looking promising for the alien hunters. There are billions of potentially habitable worlds in the Milky Way galaxy, and it is possible to interpret the early emergence of life on Earth as a hint (evidence would be too strong a word) that simple life may be inevitable, given the right conditions. The next term in the equation turns out to be more problematic for the optimist, however. We need to estimate f_i, the fraction of planets with life that go on to develop intelligent life, and f_c, the fraction of those worlds on which civilisations develop the technology to be contactable. As for the origin of life, the only evidence we have can be found in the history of life on Earth, so let us briefly summarise what we know.

The first population of living things whose ancestors survived to the present day is commonly known as LUCA – the Last Universal Common Ancestor. These four words mean something very specific; because all living things on the planet today share the same basic biochemistry, including DNA, we may assert that all living things are related and share a common origin. Specifically, if you trace your personal lineage back – to your parents, grandparents, great-grandparents and so on – you will find an unbroken line stretching all the way back to LUCA. It is possible that life emerged more than once on Earth, with different biochemistry, but we have no evidence of it. LUCA may have been unrecognisable when compared to today's life – they may not even have been cellular in nature, but rather a collection of biochemical reactions involving proteins and self-replicating molecules, possibly contained inside rocky chambers around deep-sea hydrothermal vents. They would certainly have been simpler than the earliest known microbial mats, but somewhere in your genome there will be sequences of DNA that have been faithfully passed down across the great sweep of geological time, and if you have children, you'll pass these four-billion-year-old messages on to them.

Our task is to try to estimate how likely it is that, given enough time, LUCA will evolve into organisms capable of building a civilisation. This is, of course, not precise; no accurate scientific statements can be made with a sample size of one! All we know for sure is that it happened here. The best we can do is trace our lineage back through time and try to identify potential bottlenecks along the way.

Our species, Homo sapiens, emerged around 250,000 years ago in the Great Rift Valley of East Africa. Given that Homo sapiens is the only species to have built a civilisation, the probability of our evolution from earlier hominin species is what we need to know to estimate f_c. To summarise, the emergence of Homo sapiens was undoubtedly fortuitous, dependent on many factors including, it appears, the geology of the Rift Valley itself and the details of cyclical changes in the Earth's orbit. But given enough time and the existence of large numbers of relatively intelligent animals on Earth, it is at least possible to imagine that some other creature may have made the long journey towards civilisation at some point in the future had we not emerged when we did. This is, of course, simply my opinion, and you should make up your own mind after reading further. Incredibly fortunate as we are to exist, therefore, I don't think the ascent from primates to humans is the most important evolutionary bottleneck in the road to technological civilisation, given the pre-existing biological diversity on Earth and a few tens or hundreds of millions of years of stability into the future. Rather, I think we should direct our attention back over the much longer time periods between the origin of life on Earth and the emergence of the first intelligent animals. We are mammals, which first appeared 225 million years ago in the Triassic era. Dinosaurs also appeared around this time, a subgroup of

RIFT VALLEY
Lake Natron, in the East African Rift, Tanzania. Birthplace of Homo sapiens.

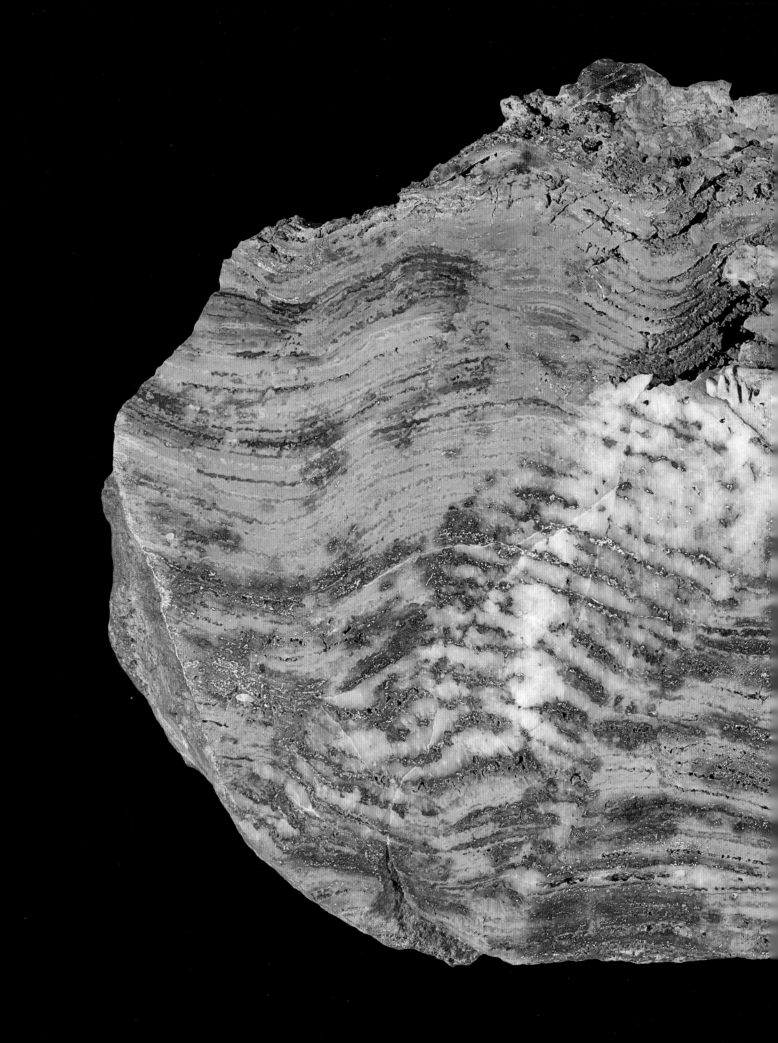

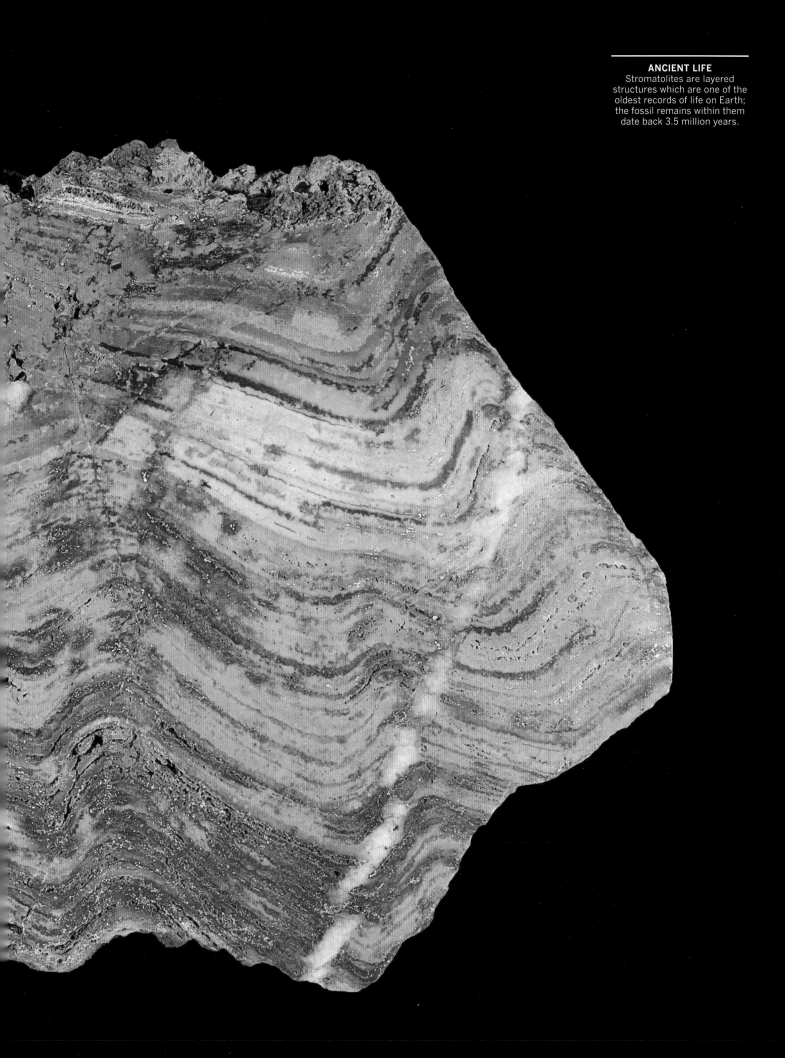

archosaurs to which birds and crocodiles are related. The first evidence of large numbers of complex animals can be found around 530 million years ago, during a period of rapid biological diversification known as the Cambrian explosion. The earliest fossils of multicellular organisms, known as Ediacaran biota, have been identified as far back as 655 million years. Many of these organisms appear sponge-like or quilted, and nothing like them survives today. There is evidence of animal-like body plans in some Ediacaran fossils, with a clearly differentiated head, but because of their soft bodies fossils are rare and relatively little is known about them. Beyond 655 million years ago, there is no evidence of multicellular life on Earth.

The half a billion years or so from the Cambrian explosion to the present day is, in geological terms, relatively short, and life seems to have marched towards greater complexity ever since. This is a gross oversimplification, and we certainly do not suggest that evolution can be viewed as an inevitable march towards intelligence. One might be tempted to assert, however, that given something akin to a Cambrian explosion, the probability of developing intelligent life may be non-negligible, although there are scientists who will strongly disagree.

There is a significantly longer stretch of time between LUCA and the Cambrian explosion – over 3 billion years – and if we are looking for potential barriers to the emergence of intelligence we should investigate the vast expanse of time before complex, multi-cellular life appeared. Why did single-celled organisms remain 'simple' on Earth for so long? Most biologists would point to at least two crucial evolutionary innovations that were necessary, though not sufficient, to trigger the Cambrian explosion. The first was oxygenic photosynthesis. An oxygen atmosphere is probably a necessary precursor for the development of

complex living things. All multicellular animals today breathe oxygen. This is not a coincidence or a biological fluke; it is chemistry. We release the stored energy from our food by oxidising it – a chemical reaction that is around 40 per cent efficient in the presence of oxygen. Food can be oxidised by other elements such as sulphur, but these reactions typically have an efficiency of 10 per cent or less. If a food chain is to be supported, with predators eating prey that eat plants and so on, then oxygen is probably essential. Without it, the energy available for predators would diminish by 90 per cent at each step in the food chain. This wouldn't simply mean that an oxygen-starved planet could be full of grazing animals like sheep and cows but no predators such as cats or sharks or humans. The arms race between predators and prey was a vital evolutionary driver towards living complexity on Earth; eyes, ears and brains offer a survival advantage whether you are the hunter or the hunted, and if predation had been impossible for energetic reasons it is far less likely or perhaps impossible that complex animals would have evolved.

Photosynthesis has been around for a long time. The 3.5-billion-year-old Western Australian microbial mat structures are bacterial and they were probably early photosynthesisers, using light from the Sun to grab electrons off hydrogen sulphide and force them onto carbon dioxide to form sugars. They would not have used a pigment as complex as the green chlorophyll that colours the landscapes of Earth today; more likely they would have used simpler molecules from the same family known as porphyrins, which occur naturally and whose precursors have been found in Moon rocks and in interstellar space. Living things are like electrical circuits – they need a flow of electrons to power their metabolism, and given the ready availability of sunlight and naturally occurring molecules that can be assembled into machines to capture it and deliver electrons, it is not too difficult to see how primitive photosynthesis might have appeared very early in the history of life on Earth.

STROMATOLITES: A RARE SIGHTING
The number of stromatolites around the world is fast declining, having fallen prey to grazing. These specimens have been preserved in the Hamelin Pool Marine Nature Reserve in Western Australia.

Given the obvious advantage of using the light from the Sun to power the processes of life, it's not surprising that some early bacteria used photosynthesis for a different purpose – to synthesise a molecule known as adenosine triphosphate, or ATP, the energy storage system for life. ATP is one of the molecules that all living things share, and must therefore be very ancient, perhaps dating back to LUCA and the origin of life.

The type of photosynthesis found in modern plants, trees and algae is a hybrid of these two processes, with an important twist. Crucially, the electrons are no longer taken from hydrogen sulphide, but from water. The fusion of these two slightly different types of photosynthesis, and the use of sunlight to grab input electrons off water, was the great evolutionary leap that led to the oxygenation of the Earth's atmosphere. Known as oxygenic photosynthesis, it evolved at some point earlier than 2.5 billion years ago. We know this because at this time Earth started to rust, forming great orange iron oxide layers known as banded iron formations, and this requires the presence of large amounts of free oxygen in the atmosphere. Molecular oxygen is an unstable and highly reactive gas, and must be constantly replenished. Astronomers in search of life on exoplanets would consider the detection of an oxygen atmosphere as a smoking gun for the presence of photosynthesis. Oxygenic photosynthesis is a terrifically complicated process, though; the molecular machinery is known as the Z-scheme, and its operation has only been understood in detail in the last few years. The sugar-manufacturing part alone, known as photosystem 2, consists of 46,630 atoms. The structure of the part that holds water molecules in place ready for their electrons to be harvested, known as the oxygen-evolving complex, was discovered in 2006. It is perhaps not surprising, therefore, that the more primitive forms of photosynthesis were not combined together into the oxygen-releasing Z-scheme for well over a billion years.

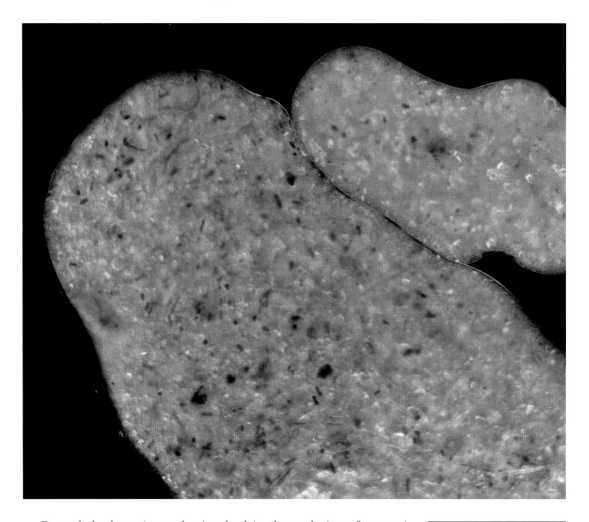

Beyond the long timescales involved in the evolution of oxygenic photosynthesis, however, there is another piece of circumstantial evidence that may suggest an evolutionary bottleneck. All the green plants and algae that fill our atmosphere with oxygen today perform their photosynthesis inside structures called chloroplasts. Chloroplasts look for all the world like free-living bacteria, and that is because they were, long ago. The story is that a bacterium, most likely one of the great family of early photosynthesisers known as cyanobacteria, was swallowed up by another cell and became co-opted to perform the complex task of grabbing electrons off water and using them to manufacture ATP and sugars, releasing the waste product oxygen in the process. This engulfing of one cell by another, and the merging of their properties, is known as endosymbiosis, an ability possessed by some cells that allows for step-changes in living things through the wholesale merger of capabilities that evolved separately and over vast periods of time in different organisms. But here is the key point; everything on the planet today that performs oxygenic photosynthesis does it using the Z-scheme, and this strongly implies that it only evolved once, most probably in a population of cyanobacteria over 2.5 billion years ago. This tremendously advantageous innovation was so useful that it became co-opted into every plant, every tree, every blade of grass and every algal bloom on the planet, flooding the atmosphere with the oxygen necessary for the Cambrian explosion to populate Earth with endless forms most beautiful. If there were ever a smoking gun for a bottleneck, this is it.

But how on earth does a cell 'learn' how to engulf another one and survive? How did endosymbiosis arise? A clue, and perhaps an even more significant bottleneck, may be found in another pre-requisite for the

ENDOSYMBIOSIS IN ACTION
Visible inside this Giant Amoeba (*Pelomyxa palustris*) are the endosymbiotic bacteria.

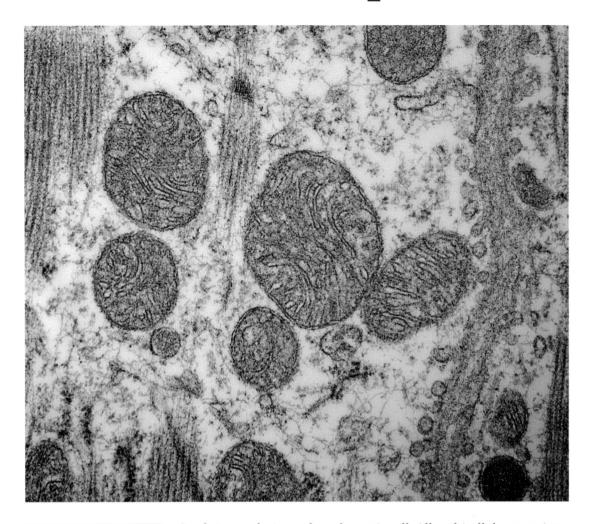

MITOCHONDRIA
These innocuous-looking circles form the power stations of cells, providing around 80 per cent of your energy through the ATP they produce.

Cambrian explosion – the eukaryotic cell. All multicellular organisms are made up of cells known as eukaryotes – cells with a nucleus and a host of specialised structures each charged with performing specific tasks. The eukaryotic cells in every living thing look so similar that an alien biologist, knowing nothing about planet Earth, would immediately recognise that human eukaryotes are closely related to those from a blade of grass. The earliest known eukaryotic cells date from around two billion years ago. Beyond this, simpler cells known as prokaryotes were the only living things on the planet. Bacteria and archaea, the two single-celled kingdoms of life that still flourish today, are prokaryotes. They are simple in the sense that they lack the vast, specialised machinery of the eukaryotes, although as we've seen they do possess some vital and extremely complex abilities – photosynthesis being a very good example.

The most striking difference between eukaryotes and prokaryotes is the eukaryotes' cell nucleus, which contains most of its DNA. In the story of evolution of life on Earth, however, it is the small amount of DNA stored outside the nucleus that is most revealing. Almost all eukaryotic cells contain structures called mitochondria. The word 'almost' is used a lot in biology. Unlike physics, there always seem to be one or two exceptions that ruin sentences in books like this. Most biologists believe that even the eukaryotes that don't possess mitochondria did so at some point in the past, however, so we can take it that these structures are ubiquitous. Mitochondria are the power stations of the cell, and their job is to produce ATP. Around 80 per cent of your energy comes from the ATP produced in mitochondria, and without them you certainly wouldn't exist. A clue as to their evolutionary origin is contained in their DNA, which is stored

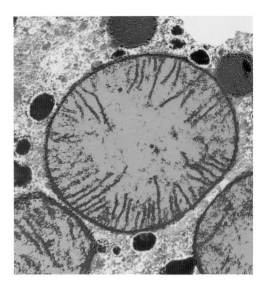 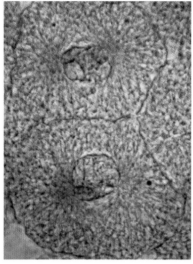 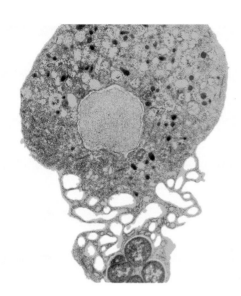

in loops and kept separate from the genetic material in the cell nucleus. Bacteria also store their DNA in loops, and this is not a coincidence. The mitochondria were once free-living bacteria.

The obvious question is, how did the bacterial mitochondria get inside the cells of every complex organism on the planet? The answer is through endosymbiosis, just as for the chloroplasts, but there is not universal agreement on the detail, and the detail matters a great deal. What is not in question is that the mitochondria are bacterial in origin. The debate surrounds the nature of the original host cell. One camp of biologists believes that the host cell was already a eukaryote, which over many millions of years had evolved an ability called phagocytosis – the ability to ingest other cells. This is a traditional Darwinian explanation – one in which complex traits evolve gradually over time via mutations and natural selection. If this is true, then it is possible to view the eukaryotic cell as just another evolutionary innovation, albeit a very important one, that might crop up anywhere given enough time. The other possibility, which is favoured by many biologists, has different implications. The idea is that the swallowing of the proto-mitochondrial cell was the origin of the eukaryotic cell itself. There was no such thing as phagocytosis or the eukaryotic cell before this singular event, and this 'fateful encounter' changed everything. Recent DNA evidence suggests that the host cell was probably an archaeon, one of the two great prokaryotic domains. Somewhere, in some primordial ocean, this simple prokaryote managed to swallow a bacterium – a trick that neither cell possessed before – and against terrific odds the pair survived and multiplied. The archaeon gained a huge advantage – a previously unimaginable energy supply from the bacterium's sophisticated ATP factory. The bacterium also gained an advantage – it was protected and, over aeons, could specialise and concentrate entirely on producing energy for its host. If this theory is correct, the origin of complex life on Earth was a complete accident. Without access to the energy supply from the mitochondria, all the complexities of the eukaryotic cell, which are absolutely necessary for complex multicellular life, would never have evolved. Earth would be a living planet today, but a planet of prokaryotes, and certainly not home to a civilisation.

I cannot tell you which of these two theories is true. If it were obvious, then all academic biologists would agree. But my impression is that the fateful encounter is currently the more widely accepted theory, and if it is correct then this has very important consequences for estimating the probability of the evolution of intelligent life. Eukaryotes are absolutely essential for intelligence. There is no biologist who

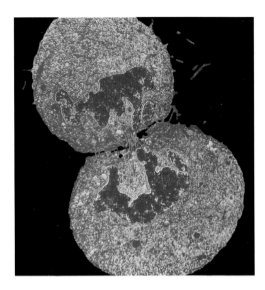 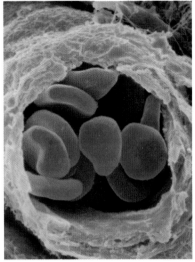 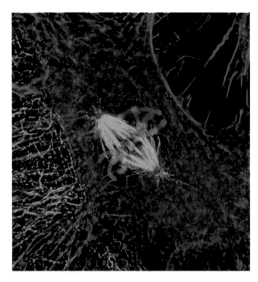

would suggest that the prokaryotes, for all their ingenuity in developing photosynthesis and mitochondrial machinery, would have managed to construct radio telescopes given enough time and a following wind. Without eukaryotes, there would be only slime.

I think these are very important points to consider in the Drake Equation. If it is correct that at least two of the necessary foundations for the emergence of complex multicellular life on Earth arose from barely credible accidents, then they might be seen as potential bottlenecks in the evolution of intelligence elsewhere in the Milky Way.

So where are we in our attempt to estimate the chances that, given the origin of life on a planet, intelligence will arise? This is where we move from science to speculation and opinion, and with these caveats, let me give you my personal view.

Given the eukaryotic cell and an oxygen atmosphere, life on Earth became diverse and complex relatively quickly. It is almost certainly no coincidence that the Cambrian explosion followed soon after a rapid rise in the oxygen content of the atmosphere. Whether it is possible to claim that intelligence on the scale necessary to build a civilisation is likely given the right biological building blocks and enough time – half a billion years, let's say – is another question. We simply don't know, and the very specific conditions in the African Rift Valley that led to the emergence of early modern humans only 250,000 years ago might suggest that civilisation-level intelligence is a rare development, even given animals as sophisticated as primates, never mind a eukaryote and an oxygen atmosphere.

An optimist would assert that there are billions of potential homes for life in the Milky Way, and that since life emerged on Earth pretty much as soon as it could at the end of the violence of the Hadean, then the Milky Way must be teeming with life and therefore civilisations. I would agree that the Milky Way must be teeming with life – I think there is a sense of chemical inevitability about it. Even accepting this line of argument, however, a pessimist would surely point to the evolution of the eukaryotic cell and oxygenic photosynthesis as being potential bottlenecks. On Earth, it took life over three billion years to get to the eve of the Cambrian. That's three billion years of planetary stability – a quarter of the age of the universe. If just one of the necessary steps – the fateful encounter, let's say – was at the fortunate end of a probability distribution, then one can easily imagine that the 20 billion Earth-like worlds in the Milky Way could all be covered in prokaryotic slime. A living galaxy, yes, but a galaxy filled with intelligence? Given what we know about the ascent from prokaryote to civilisation on Earth, I'm not so sure.

A BRIEFEST MOMENT IN TIME

Let's take one final journey back to Green Bank in 1961. Drake and his colleagues, with far less evidence than we have today, concluded that our galaxy seems remarkably conducive to life, full of Earth-like worlds warmed by the glow of benign stars. They too believed that a good fraction of these billions of worlds must be home to life, and given that Darwin's law of evolution by natural selection must apply across the universe, they concluded that intelligence must have emerged on at least some of these planets. As I've argued above, I'm not so sure about intelligence, but we must at least consider the possibility that potential evolutionary bottlenecks like the eukaryotic cell and oxygenic photosynthesis aren't as bad as they might appear. In this case, the final term in the Drake Equation

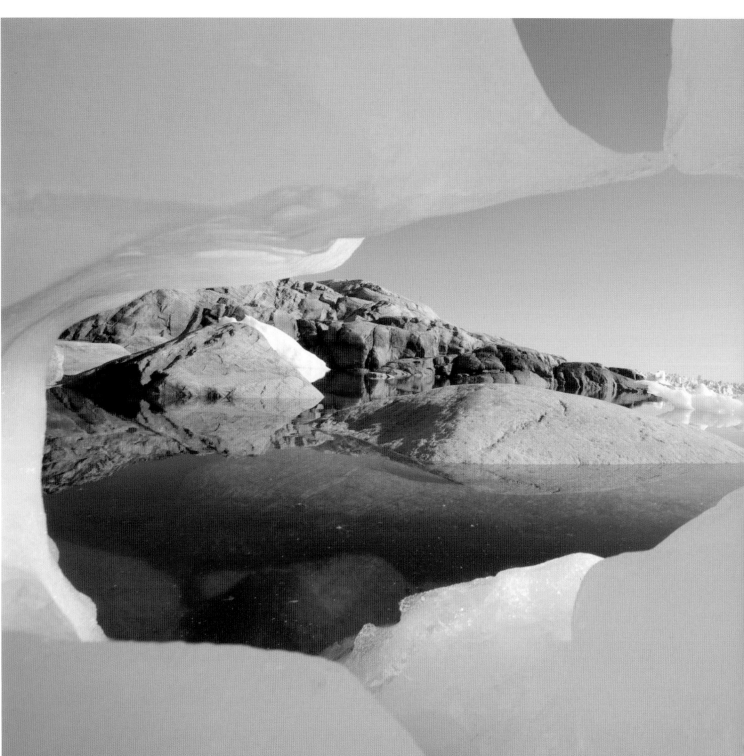

becomes all-important. Perhaps it is L, the lifetime of civilisations, that is the fundamental reason for the great silence. This is a sobering thought. The reason we have made no contact with anyone is not because of a lack of stars, or planets, or living things; it's because of the in-built and unavoidable stupidity of intelligent beings.

This might seem a bit strong, but it is a view shared at Green Bank by Manhattan Project veteran Philip Morrison. Morrison was intimately involved in the design and development of the first atomic bomb, and he helped load Little Boy onto *Enola Gay* destined for Hiroshima. The fact that human beings had deployed a potentially civilisation-destroying weapon twice, against civilian targets, and that Morrison had personally loaded one of the bombs, must have never left him, and on the eve of the Cuban missile crisis, it must have seemed likely that we would do it again on a much grander scale.

Drake realised this as well, which is certainly one of the reasons why he introduced the time that a technological civilisation can endure into his equation: we can after all only communicate with nearby civilisations if they exist at the same time as us. This is a possible resolution to the Fermi Paradox. Civilisations inevitably blow themselves up soon after acquiring radio technology, and therefore the Milky Way will remain forever silent apart from the briefest, non-overlapping flickers of intelligence. This might seem like a solipsistic conceit; how can we possibly assume that human stupidity is universal? We can't, of course. But just as for the biological arguments we made against the inevitable emergence of complex life on an otherwise living world, we only have the Earth as a guide, and extrapolating from our own experience is the best we can do. On Earth, Rutherford discovered the atomic nucleus in 1911 and we destroyed two cities and killed over 200,000 of our fellow human beings with nuclear technology 34 years later. About 17 years after that, having seen the devastation nuclear weapons can cause, Khrushchev and Kennedy came close to ending it all, and to this day we don't know how close we came to eliminating the fruits of almost four billion years of evolution. Here on Earth it appears that sanity, perspective and an appreciation of the rarity and value of civilisation emerges after, and not before, the capability to build big bombs. We have the bombs, but I don't think enough of us have the rest. Why should other young civilisations be any different? If this is the reason for the Great Silence, then I suppose we might take comfort in the fact that we are not the only idiots to have existed in the Milky Way, but that's the coldest comfort I can imagine.

The above might be seen as a naïve rant, of course. One could argue that mutually assured destruction, the guiding principle of the Cold War, did act to stabilise our civilisation and is still doing so today. Perhaps no intelligent beings will knowingly destroy their civilisation, which is what global nuclear war on Earth would surely do; after all, Kennedy and Khrushchev ultimately took this view. Similarly, one assumes that the submersion of Miami and Norwich by rising sea levels would silence the so-called climate change sceptics (I'd call them something different) and trigger a change of policy that will avert catastrophic, civilisation-threatening climate change in good time. It seems to me, however, that a small planet such as Earth cannot continue to support an expanding and flourishing civilisation without a major change in the way we view ourselves. The division into hundreds of countries whose borders and interests are defined by imagined local differences and arbitrary religious dogma, both of which are utterly irrelevant and meaningless on a galactic scale, must surely be addressed if we are to confront global problems such as mutually assured destruction, asteroid threats, climate change, pandemic disease and who knows what else, and flourish beyond the twenty-first century. The very fact that the preceding sentence sounds hopelessly utopian might provide a plausible answer to the Great Silence.

MELTING ICEBERG IN GREENLAND
The melting icecaps of Greenland are a reminder to us of the catastrophic threat we face from climate change and our wilful destruction of our planet.

SO, ARE WE ALONE?

What, then, is the range of estimates for the number of civilisations in the Milky Way, given the limited evidence we have at our disposal? During the filming of *Human Universe*, Frank Drake told me that the Green Bank meeting came up with a number of around 10,000, and he sees no reason to change that estimate. This would be wonderful, and makes the search for signals from these civilisations one of the great scientific quests of the twenty-first century. I strongly support SETI, because contact with just one alien civilisation would be the greatest discovery of all time, and it's worth the investment on that basis alone.

There is, however, one piece of evidence that might suggest a more lonely position for us on our little home world. In 1966 the mathematician and polymath John von Neumann published a series of lectures entitled 'Theory of Self Reproducing Automata' in which he analysed in great detail the possibility of constructing machines capable of building copies of themselves. Such machines exist in nature, of course – all living things do this routinely. In principle, therefore, one might imagine a sufficiently advanced civilisation building a self-replicating Von Neumann space probe and launching it out to explore the galaxy. On reaching a solar system, the probe would mine the planets, moons and asteroids, extracting the materials necessary to build one or more copies of itself. The newly minted probes would launch themselves out to neighbouring solar systems and repeat the process, spreading across the Milky Way. Even given the vast distances between the stars, computer models assuming currently envisioned rocketry technology suggest that such a strategy could result in the exploration of the entire Milky Way galaxy within a million years.

Science fiction? It certainly sounds like it, but if there is no objection in principle to the construction of a Von Neumann probe, then one has to develop an argument as to why we don't see any. The reason that this is difficult to do is due to timescales. The Milky Way has been capable of supporting life for over ten thousand million years. It is possible to envisage many millions of civilisations rising and falling over such vast expanses of time, and if only one had developed a successful Von Neumann probe, then the galaxy should be filled with its progeny; there should be at least one Von Neumann probe operating in our solar system today. Carl Sagan and the astronomer William Newman noticed a flaw in this line of argument. If the probes multiply exponentially and unchecked, then one can show that they consume the resources of the entire galaxy relatively quickly, and we'd certainly have noticed that! Or more accurately, we wouldn't be here to notice that. Sagan reasoned that this obvious risk would be sufficient to prevent any civilisation intelligent enough to build Von Neumann probes from actually doing so. They would be doomsday machines. Other astronomers have countered that it wouldn't be beyond the wit of such an advanced intellect to build in some fail-safe mechanism that guaranteed, for example, only one probe per solar system, or a finite lifetime for each probe. Others have argued that there may indeed be a Von Neumann probe operating in our solar system today, with appropriate fail-safe mechanisms installed to stop it eating everything. If such a probe were relatively small, perhaps sitting amongst the asteroids or even in the Kuiper Belt of icy comets beyond the orbit of Neptune, then we'd almost certainly be unaware of its presence.

Von Neumann probes wouldn't be the only signatures of ultra-advanced civilisations. Imagine a civilisation many millions of years ahead of us, carrying out engineering projects on galactic scale. Imagine interstellar starships or great space colonies constructed in otherwise uninhabitable solar systems. Why not? As I said at the start of this

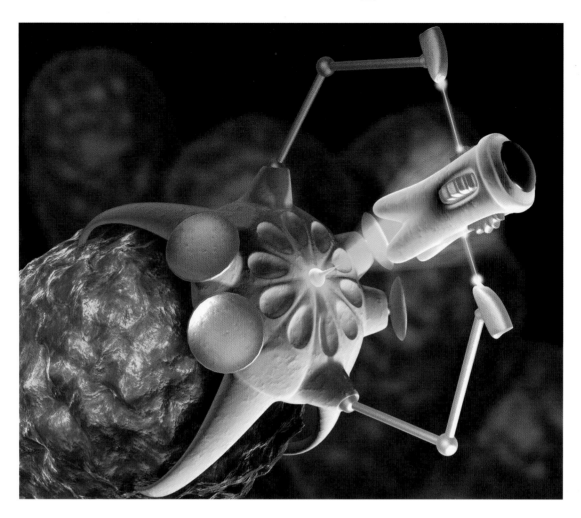

ENCOUNTERS WITH ROBOTS
Theoretically the first encounter of a space-faring civilisation is more likely to be with a self-replicating robot than with the actual life form. This computer artwork shows a nanorobot assembler using a claw to attach itself to a bacterium. One goal of nanotechnology like this is to design self-replicating systems, which would allow large manufacturing projects to become economically viable.

chapter, we went from the Wright Brothers to the Moon in a single human lifetime, so, I ask again, how far will we travel, if the laws of physics allow, given another thousand years? Or ten thousand? Or a million? What signature will we leave on the sky if we survive and prosper that long? None of these questions is trivial, because the sheer immensity of the timescales available for life to evolve in the Milky Way galaxy forces us to consider them. Why should we be the most advanced civilisation in the galaxy when we've only been building spacecraft for half a century in a 13-billion year-old universe? I don't have an answer to this. It bothers me. Perhaps the distances between the stars are indeed too great, or perhaps there are insurmountable difficulties in building self-replicating machines or starships, but I can't think what they might be.

I am tempted, therefore, to make the following argument for the purposes of debate. I think that advanced, space-faring civilisations are extremely rare, not because of astronomy, but because of biology. I think the fact that it took almost four billion years for a civilisation to appear on Earth is important. This is a third of the age of the universe, which is a very long time. Coupled with the remarkable contingency of the evolution of the eukaryotic cell and oxygenic photosynthesis – not to mention the half a billion years from the Cambrian explosion to the very recent emergence of Homo sapiens and civilisation – I think this implies that technological civilisations are stupendously rare, colossally fortuitous accidents that happen on average in much fewer than one in every two hundred billion solar systems. This is my resolution to the Fermi Paradox. We are the first civilisation to emerge in the Milky Way, and we are alone. That is my opinion, and given our cavalier disregard for our own safety, it terrifies me. What do you think?

WHO
ARE
WE?

But why, some say, the moon?
Why choose this as our goal?
And they may well ask why climb the highest mountain?
Why, 35 years ago, fly the Atlantic? ...
We choose to go to the moon.

President John F. Kennedy

SPACEMAN

Astronaut John Young was once asked how he would feel if his epitaph read 'John Young: The Ultimate Explorer'. Young smiled, and in a test pilot's drawl replied, 'I'd feel sorry for the guy who wrote it'. Young was, and still is, a hero of mine. My first vivid memory of live space exploration was watching Space Shuttle Columbia climb on a tower of bright vapour into a blue Cape sky on 12 April 1981. It was midday in Manchester, the Easter holidays, and I was 13 years old. Because of a two-day launch delay, Columbia's test flight took place precisely 20 years to the day after Yuri Gagarin made his black-and-white voyage into orbit on 12 April 1961, but Young and his co-pilot, Bob Crippen, in their orange spacesuits, were astronauts from the colour age, the future – as distant from the Russian hero as gleaming white-winged Columbia was from *Vostok 1*. Equidistant from both was Apollo, which Young flew to the Moon. Twice. It was the age of optimism, the age of wonder, the golden age when the ape went into space. When unflappable aviator Young, whose pulse rate did not increase during the launch of NASA's only manned spacecraft ever to have flown without an un-manned test flight, piloted Columbia back for a flawless manual landing at Edwards Air Force Base two days later, he turned to Crippen and said 'We're not too far away – the human race isn't – from going to the stars'.

In 2014 the stars feel further away than they did in 1981; the International Space Station is a wonderful piece of engineering that has allowed us to learn how to live and work in near-Earth orbit, but it is no closer to the stars than Columbia. Its construction is no mean achievement; one of the most important things to realise about engineering at the edge is that the only way to learn is to actually do it. You can't think your way into space; you have to fly there. But I can't help but feel, in the words of Billy Bragg, that the space race is over and we've all grown up too soon.

It was different in Gagarin's day. Nobody is born to be a spaceman. We're apes, honed by natural selection to operate in the Great Rift Valley. Gagarin's father was a carpenter and his mother was a milkmaid. Both worked on a collective farm. Gagarin's first job at the age of 16 was in a steel mill, but after showing an aptitude for flight as an air cadet he joined the military when 21 and was posted to the First Chkalovsk Air Force Pilots School in Orenburg. Rising through the ranks, he made a name for himself as a skilled and intelligent aviator, and in early 1960 he was chosen along with 19 other elite pilots for the newly established space programme. Standing just 5 foot 2 inches tall, Gagarin had the right stuff and was perfect for the tiny *Vostok* spacecraft, whose single-seat crew compartment was only 2.3m in external diameter. After a year of training, Nikolai Kamanin, head of the cosmonaut programme, chose Gagarin ahead of his rival, Gherman Titov, just four days before the flight. The history books are filled with the names of great men and women whose presence in the collective memory of humanity was assured by the slimmest of margins. Gagarin, alongside Armstrong, will be remembered for as long as there are humans in the cosmos; the name of the equally brilliant Titov, Russia's second cosmonaut, has faded away.

Gagarin's flight was a true journey into the unknown. Strapped on top of the Vostok-K rocket, which flew 13 times and made it into space on 11 occasions, the 27-year-old performed like a true test pilot. Despite a two-hour delay during which every component of the spacecraft hatch was taken apart and rebuilt while Gagarin remained strapped into his seat, his heart rate was recorded at 64 beats per minute just before launch. This is not to say that Gagarin wasn't fully aware of what he was about to do. Before boarding, Gagarin made one of the great speeches of the age.

VOSTOK'S ORBIT OF EARTH

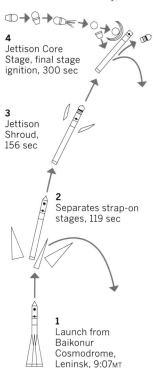

6
Begin orientation for retro burn at 800KM from landing site, 9:51MT

7
Retro burn and instrument module separation, 10:25MT. Begin re-entry, 10:35MT

4
Jettison Core Stage, final stage ignition, 300 sec

3
Jettison Shroud, 156 sec

2
Separates strap-on stages, 119 sec

1
Launch from Baikonur Cosmodrome, Leninsk, 9:07MT

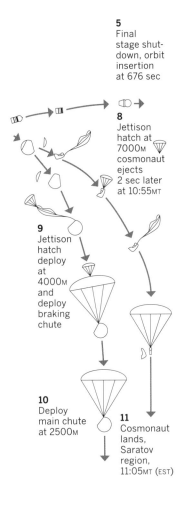

5
Final stage shutdown, orbit insertion at 676 sec

8
Jettison hatch at 7000M cosmonaut ejects 2 sec later at 10:55MT

9
Jettison hatch deploy at 4000M and deploy braking chute

10
Deploy main chute at 2500M

11
Cosmonaut lands, Saratov region, 11:05MT (EST)

THE SPACE RACE BEGINS
Vostok 1 ensured the Soviets and spaceman Yuri Gagarin secured a place in the history books. On 12 April 1961, *Vostok* and Gagarin successfully made the first manned flight into space. The race was on in space exploration.

'Dear friends, both known and unknown to me, fellow Russians, and people of all countries and continents, in a few minutes a mighty spaceship will carry me into the far-away expanses of space. What can I say to you in these last minutes before the start? At this instant, the whole of my life seems to be condensed into one wonderful moment. Everything I have experienced and done till now has been in preparation for this moment. You must realise that it is hard to express my feeling now that the test for which we have been training long and passionately is at hand. I don't have to tell you what I felt when it was suggested that I should make this flight, the first in history. Was it joy? No, it was something more than that. Pride? No, it was not just pride. I felt great happiness. To be the first to enter the cosmos, to engage single-handed in an unprecedented duel with nature – could anyone dream of anything greater than that? But immediately after that I thought of the tremendous responsibility I bore: to be the first to do what generations of people had dreamed of; to be the first to pave the way into space for mankind. This responsibility is not toward one person, not toward a few dozen, not toward a group. It is a responsibility toward all mankind – toward its present and its future. Am I happy as I set off on this space flight? Of course I'm happy. After all, in all times and epochs the greatest happiness for man has been to take part in new discoveries. It is a matter of minutes now before the start. I say to you, "Until we meet again", dear friends, just as people say to each other when setting out on a long journey. I would like very much to embrace you all, people known and unknown to me, close friends and strangers alike. See you soon!'

HERO OF HIS TIME
Yuri Gagarin's flight into space made headline news in the Soviet Union and internationally, and his name and fame have endured through history in his home country and across the world.

ТРИУМ**Ф** ЭРЬ

Весь мир восхищен беспример

It's too easy to attach trite labels to human actions – magnificent, horrific and everything in between – based on a simplified view of their causes. One can argue that the rockets carried aloft the egos of the superpowers alongside the astronauts, and this is surely right. But Gagarin spoke these words, and I challenge anyone to read them and not detect sincerity. All our actions mask a morass of motivations, worthy and less so, and the greatest human adventures are no less noble for that.

At 9.07am local time, Gagarin blasted off from Baikonur Cosmodrome in Kazakhstan, as every Russian cosmonaut has done since. Within 10 minutes, he was orbiting Earth at an altitude of 380 kilometres. His route took him across the Siberian wastes and the Pacific Ocean above the Hawaiian islands, past the tip of South America and into the South Atlantic, where he was greeted by a second sunrise before a 42-second de-orbit burn over the Angolan coast slowed *Vostok 1* into a parabolic orbit and an 8-g deceleration inside Earth's thickening atmosphere. The journey once around his home world took 1 hour and 48 minutes. Gagarin ejected from the capsule 7 kilometres above ground and, as planned, cosmonaut and spacecraft completed the final descent apart. Gliding back to Earth by parachute, Gagarin landed 280 kilometres away from the intended landing site near the Russian city of Engels. Dressed in orange spacesuit and white helmet, a farmer and his daughter bore sole witness to his historic return. 'When they saw me in my space suit and the parachute dragging alongside as I walked, they started to back away in fear,' recollected Gagarin later. 'I told them, don't be afraid, I am a Soviet citizen like you, who has descended from space and I must find a telephone to call Moscow!'

EVOLUTION OF HOMINIDS

These hominid evolutionary trees trace our genetic history as humans to the Old World monkeys that roamed Earth 25 million years ago. Discoveries of various remains, including those of the famous *Australopithecus afarensis* skeleton, commonly called Lucy, have helped us piece together an idea of our ancestry. It is believed that around 7 or 8 million years ago we split from the chimpanzees and the process of evolution onto bipedal Homo sapiens began as these monkeys started to spend more time on the ground than in the trees.

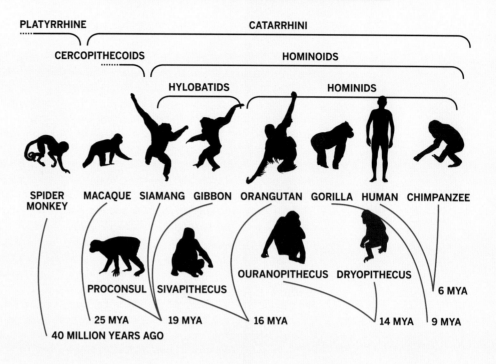

EVOLUTIONARY TREE OF MONKEYS AND PRIMATES

PLATYRRHINE — CATARRHINI

CERCOPITHECOIDS — HOMINOIDS

HYLOBATIDS — HOMINIDS

SPIDER MONKEY · MACAQUE · SIAMANG · GIBBON · ORANGUTAN · GORILLA · HUMAN · CHIMPANZEE

PROCONSUL · SIVAPITHECUS · OURANOPITHECUS · DRYOPITHECUS

40 MILLION YEARS AGO · 25 MYA · 19 MYA · 16 MYA · 14 MYA · 9 MYA · 6 MYA

HOMINID EVOLUTION

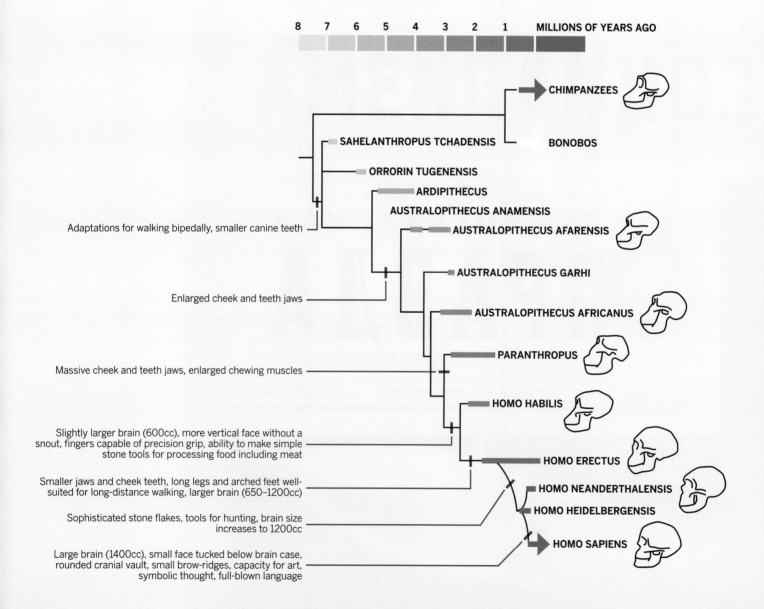

8 7 6 5 4 3 2 1   MILLIONS OF YEARS AGO

CHIMPANZEES

SAHELANTHROPUS TCHADENSIS

BONOBOS

ORRORIN TUGENENSIS

ARDIPITHECUS

AUSTRALOPITHECUS ANAMENSIS

Adaptations for walking bipedally, smaller canine teeth

AUSTRALOPITHECUS AFARENSIS

AUSTRALOPITHECUS GARHI

Enlarged cheek and teeth jaws

AUSTRALOPITHECUS AFRICANUS

PARANTHROPUS

Massive cheek and teeth jaws, enlarged chewing muscles

HOMO HABILIS

Slightly larger brain (600cc), more vertical face without a snout, fingers capable of precision grip, ability to make simple stone tools for processing food including meat

HOMO ERECTUS

Smaller jaws and cheek teeth, long legs and arched feet well-suited for long-distance walking, larger brain (650–1200cc)

HOMO NEANDERTHALENSIS

HOMO HEIDELBERGENSIS

Sophisticated stone flakes, tools for hunting, brain size increases to 1200cc

HOMO SAPIENS

Large brain (1400cc), small face tucked below brain case, rounded cranial vault, small brow-ridges, capacity for art, symbolic thought, full-blown language

APEMAN

Primates appeared relatively recently in the history of life on Earth. Studies of mitochondrial DNA suggest the Strepsirrhini suborder, containing the ancestors of Madagascar's lemurs, diverged from our own Haplorhini suborder approximately 64 million years ago, which implies that a common ancestor was present before this time, but not a great deal earlier. The first complete primate fossil found to date is that of a tree-dwelling creature known as *Archicebus achilles*, dated at 55 million years old. Discovered in the fossil beds of central China in 2013, this tiny creature would have been no bigger than a human hand, making it not only the oldest but also one of the smallest known primates.

Our family, known as the Hominidae, or more commonly the great apes, share a common ancestor with Old World monkeys around 25 million years ago, and during the making of *Human Universe* we filmed a rare species of these distant cousins in the Ethiopian Highlands. The road out of Addis towards the 3000-metre Guassa Plateau is excellent to a point, and then not excellent. The scenery, on the other hand, improves with altitude. Golden grasses illuminated by shifting lambent light through dark clouds cling to near-vertical mountainsides framing pristine villages along the high valley floors. It is fresh, cold and insect-less on the peaks above the Rift; a place to drink tea and eat *shiro*, a spiced Ethiopian stew of chickpeas and lentils. After a night in the cold but magnificently desolate Guassa community lodge, we set off at dawn to intercept the gelada baboons on their way back to their caves and ledges from early-morning foraging expeditions on the higher slopes.

The gelada baboons are a species of Old World monkey found only in the Ethiopian Highlands. They are the only surviving species of the genus Theropithecus that once thrived across Africa and into Southern Europe and India. The males in particular are powerful, long-haired animals, weighing over 20 kilograms with a bright red flash of skin on their white chests. I was told not to look them in the eye, so I didn't. Fifty thousand years ago, as our planet emerged from the last ice age, the gelada retreated into the highlands above the Rift where they still live, uniquely amongst extant primates, as graminivores, on a diet made up almost entirely of the tough high-altitude grasses and occasional herbs.

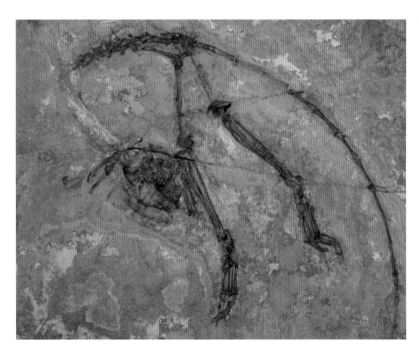

EARTH'S EARLIEST PRIMATE
Found within a rock in the Hubei Province of central China, this is the oldest complete fossil of a tree-dwelling primate found to date.

They approach with nonchalant agility in small groups, which reflect the most complex social structure of any non-human primate. Most of the groups I saw contained one or two males and perhaps eight or ten females and their young. These are referred to as reproductive units, and clearly defined hierarchies exist within them. Females usually remain in the same unit for life, but males move between them every four or five years. There are also male-only units of ten or fifteen individuals. These social units are arranged into higher groupings known as bands, herds and communities. The community we encountered numbered several hundred individuals who wandered past in their little tribes, females and young pausing to eat, groom and play whilst the larger males eyed us closely.

Despite the 25-million-year separation in evolutionary time, the gelada are very easy to anthropomorphise, especially from a vantage point amongst them, probably because their behaviour seems reminiscent of

our own and their babies are cute. Like us, they spend most of their time on the ground and operate in social groups. Some researchers familiar with the geladas claim they exhibit the most sophisticated communication behaviour of any non-human primate, employing gestures and a range of different vocalisations strung together into sequences communicating reassurance, appeasement, solicitation, aggression and defence. For all their sophistication, however, the gelada are a long way from possessing anything more complex than the simplest of human characteristics and abilities. This is, of course, an utterly obvious observation – they are monkeys! But what isn't obvious is why. The gelada separated from our common ancestor at the same time, but that self-evident statement leads to a deeper question: what is it that happened to our ancestors during those 25 million years that led us to the stars and left them on the hillsides of the Guassa Plateau eating grass?

OLD WORLD MONKEYS
The gelada baboons (*Theropithecus gelada*) are found in the grassland areas of the Ethiopian Highlands. They are the only surviving species of the genus Theropithecus that once thrived across Africa and into Southern Europe and India.

LUCY IN THE SKY

**FOOTPRINTS OF
OUR FATHERS**
Found in 1978, these footprints
are the oldest-known of early
humans. Discovered in Laetoli,
Tanzania, the 27-metre-long
footprint trail is now known as
the Laetoli Footprints.

FOOTPRINTS OF THE FUTURE
Neil Armstrong's first footprint
on the Moon on 21 July 1969, in
the Sea of Tranquility.

I am an aviation geek. I love aircraft. As I set off to film the African scenes for 'ApeMan SpaceMan', I noticed that the Ethiopian Airlines Boeing 787 I boarded at London Heathrow, bound for Addis Ababa, registration ET-AOS, was named 'Lucy'.

On the morning of 24 November 1974, Donald Johanson and a team of archaeologists were searching for bone fragments at a site near the Awash River in Ethiopia. The area was known to be a site rich in rare hominid fossils, but on that particular morning, Johanson and his graduate student Tom Gray found little to inspire them. As is often the way in science, however, a dash of serendipity coupled with an experienced scientist who understands how to increase the chances of receiving its benefits, made a seminal contribution to the understanding of human evolution. Johanson shouldn't even have been there – he had planned to spend time back at the camp updating his field notes – but as they prepared to leave, Johanson decided to wander over to a previously excavated gully and have one last look. Even though they'd surveyed the area before, this time Johanson's eye was drawn to something lying partially hidden on the slope. Closer inspection revealed it to be an arm bone and a host of other skeletal fragments – a piece of skull, a thigh bone, vertebra, ribs and jaw all emerged from the ground and, crucially, they were all part of a single female skeleton. The find triggered a three-week excavation, during which every last scrap of fossil AL 288-1 was recovered. They named it Lucy, after track 3, side one, of *Sgt. Pepper's Lonely Hearts Club Band*, because this was 1974 and they played it a lot on their tape recorder. 'Home taping kills music', they used to say back then, but it also names airliners.

Lucy lived 3.2 million years ago in the open savannah of Ethiopia's Afar Depression. Standing just over 1 metre tall and weighing less than 30 kilograms, she would have looked more ape-like than a human. Her brain was small, about one-third of the size of a modern human's and not much larger than a chimpanzee's. The anatomy of her knee, the curve of her spine and the length of her leg bones suggest that Lucy regularly walked upright on two legs, although there are a handful of scientists who would disagree. What is generally agreed upon, however, is that Lucy was a member of the extinct hominin species *Australopithecus afarensis*, and she was either one of our direct ancestors, or very closely related to them. Her bipedalism was probably an evolutionary adaptation caused by climate change in the Rift. As the number of trees reduced and the landscape became more savannah-like, the arboreal existence of our more distant ancestors became less favoured, and the increasing distances between trees selected for Australopithecus's upright gait made travel across the ground more efficient.

In 'Who Speaks for Earth?', the thirteenth chapter of Carl Sagan's *Cosmos*, there are two pictures set side by side. One is of footprints covered by volcanic ash 3.7 million years ago near Laetoli, in Tanzania, probably made by an Australopithecus afarensis like Lucy. Some 400,000 kilometres away and 3.7 million years later, another hominin footprint was left in the dust of the Sea of Tranquility. Together, they speak eloquently of our unlikely, magnificent ascent from the Rift Valley to the stars. The remainder of this chapter deals with the 3 million years between Lucy and the Moon. The timescale is ridiculously small; less than a tenth of one per cent of the period of time during which life has existed on Earth. Lucy was little more than an upright chimpanzee; an animal, a genetic survival machine. We bring art, science, literature and meaning to the Earth; we are a world away, and yet separated by the blink of an eye. 'Our obligation to survive is owed not just to ourselves but also to that Cosmos, ancient and vast, from which we spring,' wrote Sagan. I'd like to add that we owe it to Lucy as well.

FROM THE NORTH STAR TO THE STARS

Before astrology was consigned to the status of trifling funfair entertainment by science, it was believed that the position of the planets against the distant stars had a profound effect on people's daily lives. If you don't know what the stars or planets actually are, this is at least within the bounds of reason, but as our understanding of physics improved, so it became clear that there is no way that the position of a distant planet relative to the fixed stars can have any effect on the behaviour of a human being on the surface of the Earth. The planets can and do affect the Earth's motion through the Solar System over timescales far greater than those of human lifetimes, though, and recent research suggests that long-term changes in Earth's orientation and orbit may have played a crucial role in hominid evolution.

Polaris is a true giant, almost 50 times the diameter of our sun. It is also a Cepheid variable, one of the valuable standard candles upon which the astronomical distance scale rests. At a distance of only 434 light years, it is both the closest Cepheid and one of the brighter stars in the sky, dominating the constellation Ursa Minor. Polaris also happens to be aligned directly with the Earth's spin axis, and this special position on the celestial North Pole makes it invaluable to navigators. As the Earth spins on its axis, Polaris sits serenely as all other stars rotate around it. At any point in the northern hemisphere, your latitude is the angle between Polaris and the horizon; zero degrees north at the Equator, where Polaris is on the horizon, and 90 degrees north at the Pole where Polaris is directly overhead. As viewed from Oldham, Lancashire, UK, Polaris sits at an angle of 53.54 degrees above the horizon.

Christopher Columbus and Ferdinand Magellan relied on Polaris as they crossed the oceans and explored new worlds. Perhaps more surprisingly, on board Apollo 8 Jim Lovell carried a sextant as a back-up navigational device. Designed by the MIT instrument laboratory in Cambridge, Massachusetts, it may not have looked traditional but it operated in exactly the same manner as the one constructed by instrument maker John Bird in 1757. Polaris was one of Apollo's key navigational stars. It was paired with Gamma Cassiopeia on Lovell's charts, which was known in Apollo jargon as 'Navi'. The name was coined by Gus Grissom on Apollo 1 as a prank – it was his middle name 'Ivan' backwards. Two other navigational stars, Gamma Velorum and Iota Ursa Major, were named 'Regor' after Roger Chaffee and 'Dnoces' after Ed White the 'Second'. Using the stars for navigation might seem hopelessly old-fashioned, but if you think about it for a moment, you'll realise that there is no other way that a spacecraft in deep space can orient itself, other than relative to the fixed stars on the celestial sphere.

A spacecraft will often shift its position relative to the stars, but on Earth, things feel different because our orbit around the Sun is relatively stable from year to year. There are wobbles on relatively short timescales associated with changes in the speed of Earth's rotation, and these lead to the insertion of leap seconds to keep our atomic clocks synchronised with the heavens. Between 1972 and 1979, nine leap seconds had to be inserted, whilst none was needed between the beginning of 1999 and the end of 2005. Earth's rotation rate is noticeably chaotic when compared to the accuracy of atomic clocks.

The largest short-term contribution to changes in Earth's rotation comes from the gravitational influence of the Moon, which acts to slow down the rate of spin by around 2.3 milliseconds per century due to friction between the tidal bulges in the oceans and the rotating solid Earth beneath, but there are also longer-term changes. The most pronounced of

NAVIGATING THE STARS
Frank Borman, spacecraft commander on the Apollo 8 mission, relied on the latest technology to explore new worlds, but knowing that colleague Jim Lovell had brought his sextant with him if technology failed them!

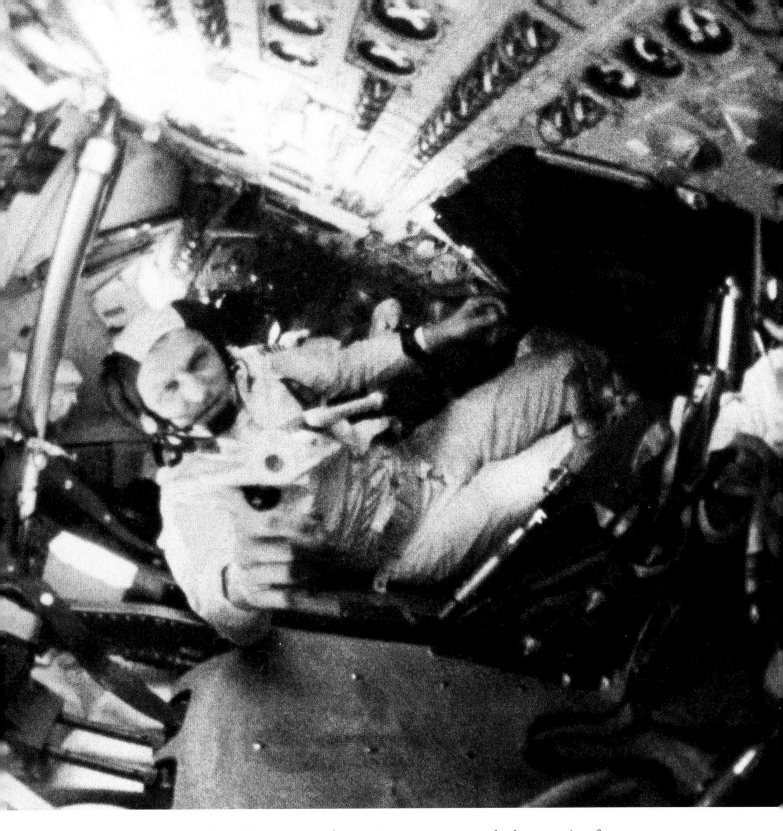

these is known as axial precession or, more commonly, the precession of the equinoxes. The Earth spins on its axis like a gyroscope, and because it spins, it bulges out at the Equator. Because the Earth isn't a perfect sphere, the gravitational influence of the Sun and Moon exerts a torque on the Earth that causes its spin axis to sweep around in a circle once every 26,000 years. This is not subtle, because the spin axis itself is tilted at 23 degrees relative to the plane of Earth's orbit, and precession therefore has a large effect on the night sky that was first documented by the Greek astronomer Hipparchus, around 150 BCE. Precession manifests itself as a shift in the position of the celestial pole relative to the fixed stars. There will come a time in the not too distant future when Polaris will no longer sit above the celestial North Pole as our spin axis traces out a circle in the sky. In about 3000 years' time, navigators of the future will rely on Gamma Cephei as a back-up for their GPS systems as they sail across the seas of our planet, and in 8000 years it will be the bright star Deneb. The identity of the North Star has altered many times throughout human history. As the Egyptians finished building the Great Pyramid of Giza in 2560 BCE, Alpha Draconis lay closest to the celestial pole. Two and a half thousand years later, as the Romans did things for us, Kochab, the second-brightest star in Ursa Minor, and its neighbour Pherkad were known as the 'Guardians of the Pole'. Precession therefore affects navigation, but more importantly it also affects our climate.

The 23-degree tilt of Earth's spin axis is responsible for the seasons; summer in the northern hemisphere occurs when the North Pole is tilted towards the Sun, leading to constant daylight within the Arctic Circle. Half a year later and the geometry is reversed, with the South Pole receiving 24-hour daylight and the southern hemisphere experiencing summer. Precession alone would have no effect on the climate if the Earth's orbit were a perfect circle, but it isn't; it is elliptical, with the Sun at one focus. At the turn of the twenty-first century, it happens to be the case that the Earth is at its closest approach to the Sun (known as perihelion) in January, just after the winter solstice when the North Pole is pointing away from the Sun. This makes northern winters slightly milder than they would otherwise be, because the Earth receives a little bit more solar radiation during the northern winter. In around 10,000 years' time, however, precession will have carried the Earth's spin axis around by a half-turn, and it will be the North Pole that points towards the Sun at perihelion, making northern hemisphere summers slightly warmer and winters cooler. The more elliptical the Earth's orbit, the more pronounced this effect.

This is where things get a little more complicated, but it's the complication that matters for our story. The planets are significantly further away than the Moon, but also significantly more massive, and their constantly shifting positions induce periodic changes to our orbit over long timescales. Jupiter has the most pronounced effect due to its large mass and relative proximity. The largest of these changes occurs on a timescale of 400,000 years. Picture the Earth's orbit becoming periodically more elliptical and more circular, stretching back and forth with a period of 400,000 years. This oscillation modulates the effect of precession on the climate; at the times when the Earth's orbit is at its most elliptical, the changes due to precession will be at their most pronounced. This effect is known as astronomical or orbital forcing of the climate.

There are many such resonances in Earth's orbit – another important change in the eccentricity of the ellipse occurs every 100,000 years. Furthermore, the tilt of the axis itself swings back and forth between around 22 and 25 degrees on a 41,000-year cycle. The whole solar system is like a giant bell, ringing with many hundreds of harmonics driven by the gravitational interactions between the Sun, planets and moons.

Over many thousands of years, these shifts in the Earth's orbit and orientation relative to the Sun have led to dramatic changes in climate, and are certainly one of the key mechanisms that drive the Earth into and out of ice ages. It is perhaps obvious that these long-term shifts in climate should have had an effect on the evolution of life; ice ages present a significant challenge to animals and plants and this will provoke an evolutionary response via natural selection. More surprisingly, recent research has suggested a direct link between precession, the 400,000-year eccentricity cycle, and the evolution of early modern humans.

ASTRONOMICAL SEASONS

The Milankovitch theory describes the collective effects of changes in the Earth's movements upon its climate. They are named after Serbian geophysicist and astronomer Milutin Milankovitch, who worked on it during his internment as a prisoner in World War One. Milankovitch mathematically theorised that variations in eccentricity, axial tilt and precession of the Earth's orbit determined climatic patterns on Earth. The Earth's axis completes one full cycle of precession approximately every 26,000 years. At the same time, the elliptical orbit rotates over a much longer timescale. The combined effect of the two precessions leads to a 21,000-year period between the astronomical seasons and the orbit. In addition, the angle between Earth's rotational axis and the normal to the plane of its orbit (obliquity) oscillates between 22.1 and 24.5 degrees on a 41,000-year cycle. It is currently 23.44 degrees and decreasing.

THE PRECESSION OF EARTH'S SPIN AXIS

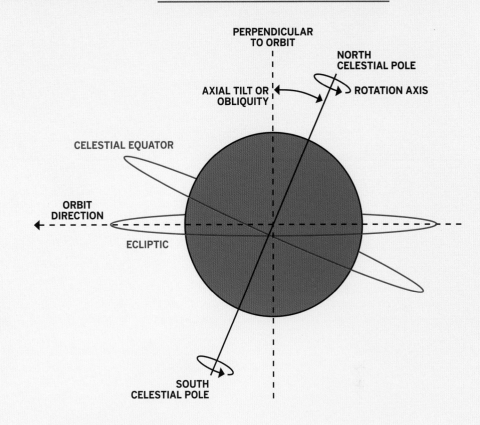

THE PRECESSION OF THE EQUINOXES

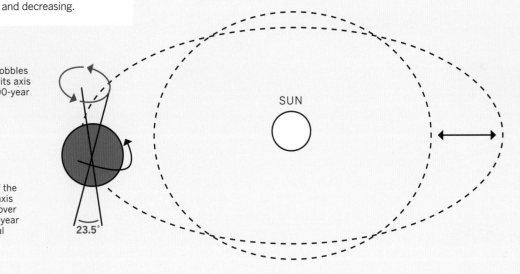

The Earth wobbles like a top on its axis over a 20,000-year cycle

The tilt of the Earth's axis changes over a 40,000-year interval

23.5°

SUN

The shape of its orbit changes the Earth's distance from the Sun over a period of 100,000 years

MILANKOVITCH CYCLES

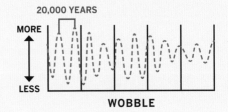

20,000 YEARS

MORE

LESS

WOBBLE

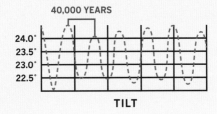

40,000 YEARS

24.0°
23.5°
23.0°
22.5°

TILT

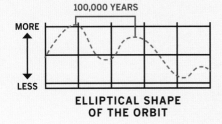

100,000 YEARS

MORE

LESS

ELLIPTICAL SHAPE OF THE ORBIT

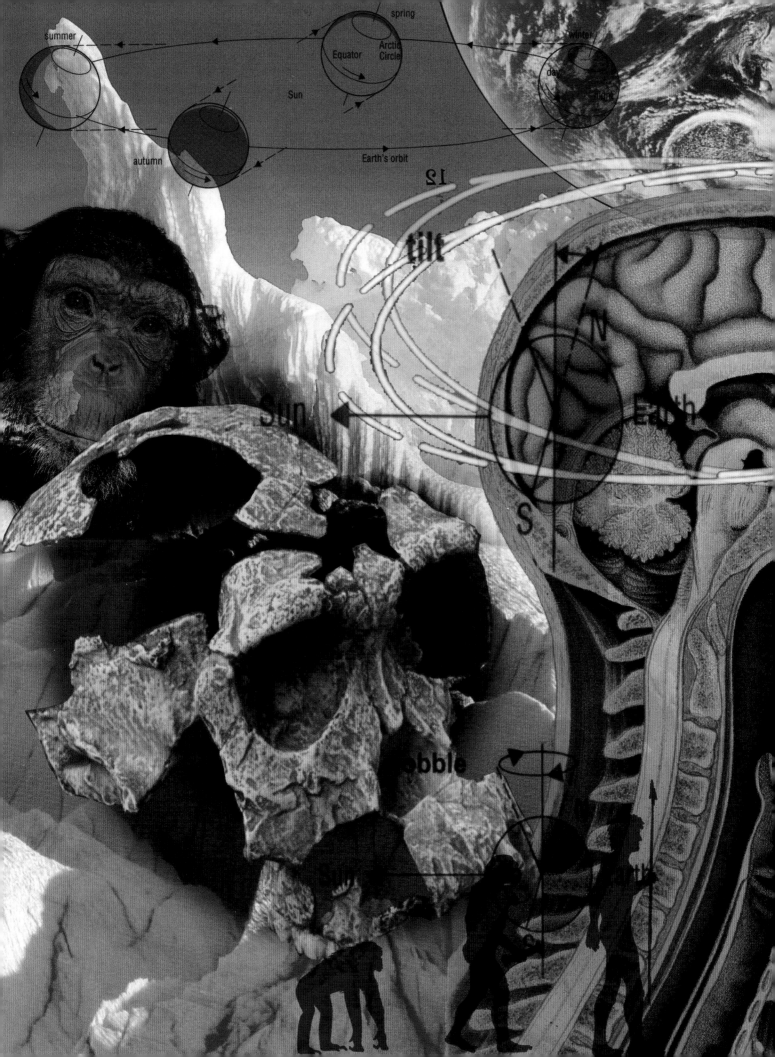

summer

spring

Equator

Arctic
Circle

winter

day

night

Sun

autumn

Earth's orbit

12

tilt

Sun

N

Earth

S

wobble

Sun

Earth

Sun

CLIMATE CHANGE IN THE RIFT VALLEY
AND HUMAN EVOLUTION

The Great Rift Valley; evocative words that immediately suggest origins. There are many reasons I love visiting Ethiopia. I love the people. I love the food. I love the high-altitude freshness of Addis. I love the mountains and valleys and high plains. I even loved visiting Erta Ale, the legendary shield volcano at the Afar Triple Junction known as the gateway to hell, although I probably won't do it again. But I also love an idea. It's impossible to visit this ancient country and not catch a glimpse in your peripheral vision of a chain of ghosts stretching back ten thousand generations, because it is

firmly embedded in popular culture that we came from here. Every one of us is related to someone who lived in Ethiopia hundreds of thousands of years ago. It is the Garden of Eden, the place where humanity began. What popular culture has yet to assimilate, however, is the fortuitous and precarious nature of the ascent of man. When I was growing up I remember talk of 'the missing link', that elusive fossil that would tie us definitively to our ape-like ancestors. When I started school, DNA sequencing was not yet invented, and Lucy hadn't been unearthed. Today, we have a significantly more complete view of how Australopithecines like Lucy are related to modern humans, and whilst the details are still debated and new evidence is continually updating the standard model of hominin evolution, it is now possible to tell the broad sweep of the story in some detail.

GATEWAY TO HELL?
The Erta Ale basaltic shield volcano is located in the Danakil depression in Ethiopia. It has an active lava lake around 100 metres in diameter. This lava lake is part of the summit caldera; the constant motion of the lava lake's surface resembles plate tectonics on a smaller scale.

CRANIAL CAPACITY

The internal volume of the primate skull increases from 275–500cc in chimpanzees to 1130–1260cc in modern humans. Neanderthals had a brain capacity in the range 1500–1800cc – the largest of any hominids. Recent research indicates that, in primates, whole brain size is a better measure of cognitive abilities.

CHIMPANZEE SKULL

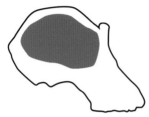

AUSTRALOPITHECINE SKULL

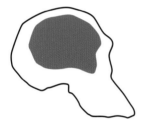

HUMAN SKULL

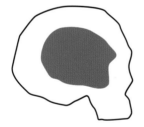

The members of our human evolutionary family are referred to as hominins. The split between hominins and the ancestor of the chimpanzee occurred at some point before 5 million years ago in Africa, and by 4 million years ago, Australopithecus afarensis – Lucy – was present. Their brain size was approximately 500cc, around the same as a chimpanzee and less than one-third of that of a modern human. Around 1.8 million years ago, there was a step change in both brain size and the number of hominin species in the East African Rift. Several species of our genus Homo appeared, including Homo habilis and Homo erectus. They lived for a time alongside other species, including several Australopithecines and a genus known as Paranthropus. There are anthropologists who prefer to classify Paranthropus as a different species of Australopithecus. I make this point not to be confusing, but to highlight an important fact; the study of hominin evolution is a difficult area, and it is not surprising that there are ongoing debates about the classification of 2-million-year-old fossils and DNA sequences. What is important for our story, however, and what nobody disputes, is that there seems to have been a jump in both brain size and the number of species of hominins in the Rift Valley region around 1.8 million years ago. By around 1.4 million years ago, only one of these species had survived – Homo erectus – with a brain size of 1000cc. The next milestone is the appearance of Homo heidelbergensis around 800,000 years ago. Homo heidelbergensis is generally accepted to be the ancestor of Homo sapiens and the Neanderthals who lived alongside us in Europe until around 45,000 years ago, and possibly later. Homo heidelbergensis represented another jump in brain size, up to around 1400cc, which is close to that of modern humans.

In the late 1960s and early 1970s, two hominin skulls were found near the Omo River in Ethiopia. Known as Omo 1 and Omo 2, argon dating of the volcanic sediments around the level they were found dates them at 195,000+/-5000 years old. These are the earliest fossilised remains to have been identified as Homo sapiens.

The interesting question is what caused these rapid increases in brain size, driving hominin intelligence from the chimpanzee-like capabilities of Australopithecus to modern humans in only a few million years. Again, this is a very active area of research, and there are differences of opinion amongst experts. This is the nature of science at the frontier of knowledge, and this is what makes science exciting and successful. The model we are focusing on is the most widely accepted theory of human evolution. It is known as the recent single origin hypothesis, or more colloquially the 'Out of Africa' model, and the dates and locations we have described so far might be referred to as 'text book'. There is broad consensus, therefore, about the 'When?' and the 'Where?'. But not 'Why?', and it is to 'Why?' that we now turn.

The figure on page 143 is reproduced from a paper published in 2013 by Shultz and Maslin. The lower figure shows the cranial capacity, or brain size, of skulls found in the Rift Valley, plotted against their age. The skulls are labelled according to their species. There is a trend towards larger brain size over the 4 million years since the emergence of Australopithecus, but the trend is not gradual. As we noted above, there is a large jump around 1.8 million years ago with the emergence of Homo erectus, and another jump just under 1 million years ago with Homo heidelbergensis. The final jump occurs when Homo sapiens emerges 200,000 years ago. The time period around 1.8 million years ago also corresponds to a leap in the number of hominin species present in the Rift Valley; there were at least five or six species living side by side, suggesting that something interesting occurred around this time which may have been responsible for, or was a contributing factor to, the observed increase in brain size, particularly in Homo erectus. A suggestion as to what this might have been can be seen in the top figure, which shows a measure

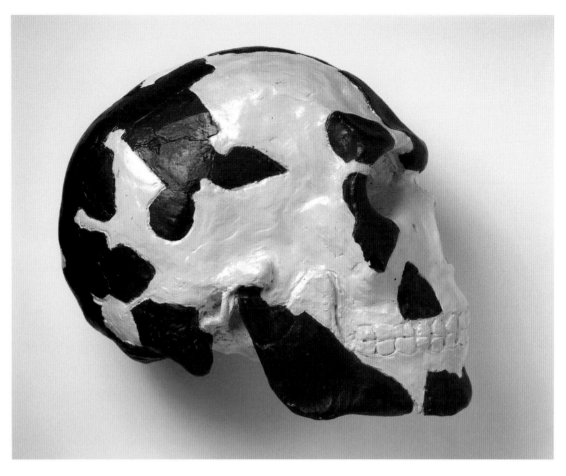

of the rate of appearance and disappearance of deep-water lakes in the Rift Valley. The large number of lakes appearing temporarily around 1.8 million years ago indicates that at this time the climate, and in particular the level of rainfall, was varying quickly and violently. Similar climate variation occurs around 1 million years ago and 200,000 years ago, and this appears to be correlated with increase in hominin brain size. The theory is that rapidly changing climatic conditions in the Rift Valley at these specific times played an important role in driving the increases in brain size. The selection pressures that may have led to these increases are unclear. Selection for adaptability was probably an important factor, but social factors such as the ability to live in large groups, and intra-species competition as a result of the larger number of species living side by side, particularly around 1.8 million years ago, must also have played a role. Having said that, it does appear that climate variation in the Rift Valley 1.8 million, 1 million and 200,000 years ago could have been a contributing factor to the development of our intelligence. This is known as the Pulse Climate Variability hypothesis.

We can now bring all these threads together to reveal a surprising and, for me, dizzying hypothesis which, if correct, sheds new light on the immensely contingent nature of the existence of our modern civilisation – or, in simpler language, why we are bloody lucky to be here!

The three dates – 1.8 million, 1 million and 200,000 years ago – correspond to the times when the Earth's orbit was at its most elliptical. As described above, the mechanism by which climate changes due to precession at these times is well understood. The Pulse Climate Variability hypothesis asserts that the unique geology and position of the Great Rift Valley amplified these changes, and that early hominins responded by increasing their brain size. If this is correct, our brains evolved as a response to changes in the Earth's orbit, driven by the precise

TRACES OF HUMAN PAST
Omo 1 and Omo 2 are the earliest fossilised remains of Homo sapiens so far recovered. The partial skulls will help scientists in their quest to uncover the secrets of our evolution.

arrangement of the orbits of the other planets around the Sun, and precession, driven primarily by the gravitational interaction between the Moon and Earth's axial tilt, both of which date back to a collision early in the history of the solar system, and all this is plainly blind luck. Without an inconceivably unlikely set of coincidences, and the way these conspired together to change the climate in one system of valleys in wonderful Ethiopia, we wouldn't exist.

If this is correct, then what a response! I held a brain for the cameras at St Paul's teaching hospital in Addis. It is the most complex single object in the known universe, a most intricate example of emergent complexity assembled over 4 billion years by natural selection operating within the constraints placed upon it by the laws of physics and the particular biochemistry of life on Earth. It contains around 85 billion individual neurons, which is of the same order as the number of stars in an average galaxy. But that doesn't begin to describe its complexity. Each neuron is thought to make between 10,000 and 100,000 connections to other neurons, making the brain a computer way beyond anything our current technology can simulate. When we do manage to simulate one, I have no doubt that sentience will emerge; consciousness is not magic, it is an emergent property consistent with the known laws of nature. But that doesn't lessen the wonder one iota. Out of this evolutionary marvel, we emerge. Thoughts, feelings, hopes and dreams exist on Earth because of electrical activity inside a 1.5-kilogram blob of stuff, which hasn't changed much since the earliest modern humans began the long journey out of Africa. If you could travel back in time and bring a newborn baby from 200,000 years ago into the twenty-first century, allowing it to grow up in our modern society with a modern education, it could achieve anything a modern child could. It could even become an astronaut. Which sets up one more question: if the hardware was present 200,000 years ago, then what changed to lift us from the Great Rift Valley into space?

CRANIOFACIAL DEVELOPMENT

Homo neanderthalensis has a unique combination of features on its skull that is distinct from fossil and extant 'anatomically modern' modern humans. Modern research involving morphological evidence, direct isotopic dates and fossil mitochondrial DNA from three Neanderthals indicates that the Neanderthals were a separate evolutionary lineage for at least 500,000 years. However, it is unknown when and how Neanderthal's unique craniofacial features emerged.

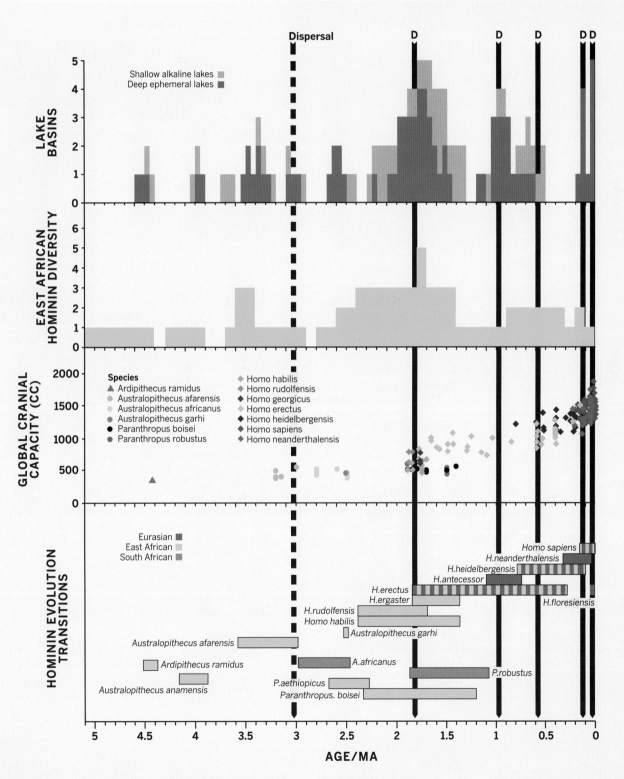

Dispersal D D D D D

LAKE BASINS

Shallow alkaline lakes
Deep ephemeral lakes

5
4
3
2
1
0

EAST AFRICAN HOMININ DIVERSITY

6
5
4
3
2
1
0

GLOBAL CRANIAL CAPACITY (CC)

Species
- ▲ Ardipithecus ramidus
- ● Australopithecus afarensis
- ● Australopithecus africanus
- ● Australopithecus garhi
- ● Paranthropus boisei
- ● Paranthropus robustus
- ◆ Homo habilis
- ◆ Homo rudolfensis
- ◆ Homo georgicus
- ◆ Homo erectus
- ◆ Homo heidelbergensis
- ◆ Homo sapiens
- ◆ Homo neanderthalensis

2000
1500
1000
500
0

HOMININ EVOLUTION TRANSITIONS

Eurasian
East African
South African

Homo sapiens
H.neanderthalensis
H.heidelbergensis
H.antecessor
H.erectus
H.ergaster
H.floresiensis
H.rudolfensis
Homo habilis
Australopithecus garhi
Australopithecus afarensis
Ardipithecus ramidus
A.africanus
P.robustus
P.aethiopicus
Australopithecus anamensis
Paranthropus. boisei

5   4.5   4   3.5   3   2.5   2   1.5   1   0.5   0

AGE/MA

'AN UNPRECEDENTED DUEL WITH NATURE'

'The best thing we can do now is just to listen and hope', said Cliff Michelmore, broadcasting from the BBC's studios 24 minutes from the expected splashdown of Apollo 13. On 17 April 1970, I was too young to watch the live broadcast, but I've seen the recording many times since. Grainy pictures from the deck of the USS *Iwo Jima*, its flight deck crammed with nervous sailors off the coast of Samoa; Patrick Moore and Geoffery Pardoe grim-faced in the studios, and James Burke, famously, with fingers crossed behind his back. 'Apollo control, Houston, we've just had loss of signal from Honeysuckle'. Honeysuckle Creek Tracking Station in Canberra, Australia, was the last ground station to contact Apollo 13 before it entered the Earth's atmosphere on its way home. Signal loss during re-entry is routine high drama on all space missions; the ionisation of the atmosphere caused by the frictional heating of the spacecraft blocks radio signals, typically resulting in radio silence for four minutes. On Apollo 13, six minutes passed in silence. The brilliance of the BBC's quartet of commentators was in the silence they allowed on the airwaves. The only sound was the static of the NASA feed – a moment of genuine tension. No need for vacuous media babble; nobody could bring themselves to speak. 'We'll only know whether that heat shield was damaged by that explosion three days ago when they come out of radio blackout in just over two minutes' said Burke. Silence. As four minutes passed, Houston reports '10 seconds to end of radio blackout'. Silence. Houston: 'We've had a report that Orion 4 aircraft has acquisition of signal.' 'They're through' says Burke. 'Let's not anticipate, because the parachutes may have been damaged.' 'Shutes should be out', murmurs Burke; not broadcasting, just saying. 'There they are, there they are!' 'They've made it' remarks Moore. And then applause. 'I make it no more than 5 seconds late!' shouts Burke, 'No more than 5 seconds late!'

The safe return of Apollo 13 was arguably NASA's finest hour; 55 hours 54 minutes and 53 seconds into the mission, 320,000 kilometres from Earth, Lunar Module pilot Jack Swigert switched on a system of stirring fans in the hydrogen and oxygen tanks in the service module, a routine procedure. A piece of Teflon insulation inside the tank had been damaged, it was later discovered, by a series of unlikely events that happened on the ground during the preparation of the spacecraft for flight. The wire shorted, the tank exploded, and the side of the service module was blown off, critically damaging the spacecraft's power supply systems and venting the crew's oxygen supply out into space.

The Command Module, the only part of the spacecraft capable of surviving a re-entry through the Earth's atmosphere, was now running on batteries and with a rapidly diminishing oxygen supply that would not keep the astronauts alive long enough to return to Earth. The only option was to shut down the Command Module and retreat to the Lunar Module, effectively using it as a life raft. Lovell later spoke of how he didn't regret the mission at all. He was robbed of his Moon landing, which must have been doubly frustrating given he'd already flown to the Moon on the historic Apollo 8 mission. But his reaction, revealed in interviews in later life, offers great insight into the character of a test pilot. 'We were given the situation,' Lovell explained, 'to really exercise our skills, and our talents to take a situation which was almost certainly catastrophic, and come home safely. That's why I thought that 13, of all the flights – including [Apollo] 11 – that 13 exemplified a real test pilot's flight.' Both Lovell and Haise have said that the idea of not returning safely to Earth never really came up. 'There was nothing there that said irrefutably we don't have a chance'.

Haise was correct, of course, because they did return safely. But they only had enough food and water to sustain two people for a day and a half

THE RETURN OF APOLLO 13
Apollo 13 was launched on 11 April 1970, and was planned as the third manned Moon landing. Two days into the flight, 300,000 kilometres from Earth, an oxygen tank in the spacecraft exploded. With their normal supplies of oxygen and electricity destroyed, the astronauts used the Lunar Module as a 'life raft' to keep them alive.

and had to improvise a carbon dioxide filter to provide them with enough breathable air for the return journey. Locked in the Lunar Module with limited supplies of food and water and temperatures dropping towards freezing, life was far from comfortable. With the Command Module powered down to preserve the sparse battery supplies left after the loss of the fuel cells, the crew had to survive in a hostile environment with limited resources. Like so many outposts of human civilisation throughout history, shortage of water was a primary concern. Water was critical on the Lunar Module for two reasons; as well as being needed to keep the crew hydrated and to rehydrate the food, it also cooled the electrical systems on the spacecraft. Conserving water therefore became a critical part of the plan to return to Earth. Reducing their intake to just one-fifth of a normal human water ration, each of the crew suffered severe dehydration and together they lost 31.5 pounds in weight – nearly fifty per cent more than any other Apollo crew.

Despite the discomfort, setting a new mission trajectory and navigating their way along it remained the primary challenge. The standard way to make in-flight course corrections on Apollo was to use the Command Module's main engine, but the system was located close to the damaged site and mission controllers decided that lighting it was too great a risk. Instead, the decision was made to use the LM's descent engine to send them around the far side of the Moon and back to Earth in four and a half days. This is known as a free-return trajectory – a slingshot around the Moon at the correct angle to return directly to Earth. No one knew if an engine designed for a completely different purpose would perform this function successfully – but they knew that if it failed they would not return.

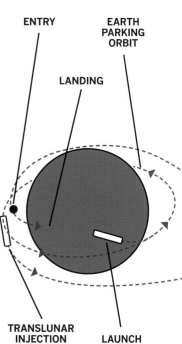

HOW APOLLO 13 GOT BACK

ENTRY · EARTH PARKING ORBIT · LANDING · TRANSLUNAR INJECTION · LAUNCH

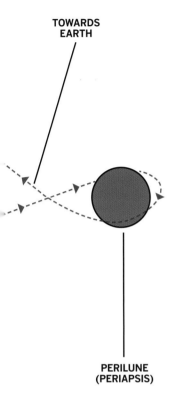

TOWARDS
EARTH

PERILUNE
(PERIAPSIS)

Five hours after the initial explosion, the LM engine was fired for a 35-second burn, successfully putting the crew onto a free return trajectory. This solved one problem but raised another. Calculations of the trajectory estimated return to Earth 153 hours after launch, which would push the key reserves on the craft too low for comfort, so it was decided to speed up the spacecraft with another burn, cutting the total time of the voyage by ten crucial hours. Such were the slim margins on Apollo 13. The main navigation system in the Command Module was out of action, so Lovell had to calculate the correct navigational inputs, while back at base, mission control worked through the same calculations as a cross check. Lovell also got to use his sextant, which he played with on Apollo 8, to navigate by the stars for real.

The calculations are preserved as handwritten notes, in the Lunar Module System's Activation Checklist. This was the checklist Lovell and Haise would have used to fly down to the Moon's surface. Now useless, Lovell used the waste paper to write down instructions to put the ship on course for Earth. Two hours after they rounded the far side of the Moon, the LM engine fired, following Lovell's handwritten checklist, increasing the speed of the spacecraft by 860 feet per second and buying them ten precious hours.

The most dramatic rescue in the history of human spaceflight stands as a testament to the brilliance of the three test pilots Lovell, Haise and Swigert, and also to the brilliance of the engineers on the ground who simply knew their stuff. NASA's Apollo engineers were young by today's standards; the average age of the team in mission control for the splashdown of Apollo 11 was 28 years old. This is one of the reasons

NOTES FROM SPACE
Astronaut Jim Lovell's handwritten checklist which helped ensure the Apollo 13 crew's safe return to Earth.

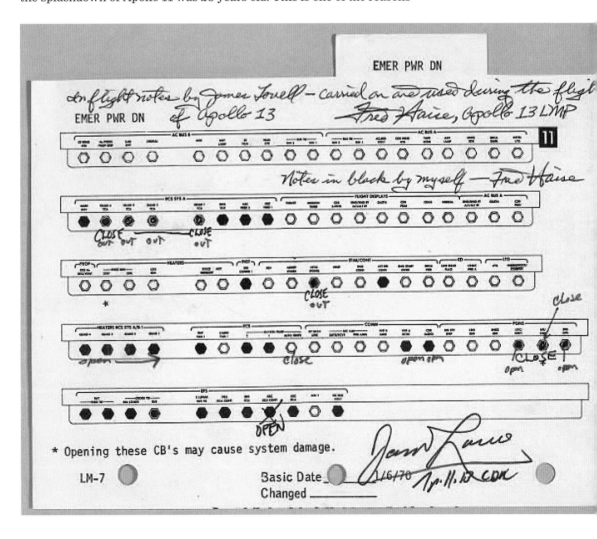

ACT 30

7:26		97:26

IMU COARSE ALIGN

CSM In Min DEADBAND ATT HOLD

te LM Gimbal Angles

298
360 / 8
658

	IG	MG
0	180.00	360.00

+Rc (See TLC-1)

-CM 167.78 +CM 351.87 -CM
(112.5)   (22.5)

LM 347.78 LM 208.13 LM
(292.5)   (337.5)

E COARSE ALIGN IMU
LOAD ICDU ANGLES OG,IG,MG (.01°)
T Lt - On, FDAI Torques)
*PROG Lt-On   *
*V05 N09E 00211 COARSE *
*   ALIGN ERROR,Go*
*   To 3   *

SUIT FAN/H2O SEP CHECK

1 CB(16) ECS: SUIT FAN 2 - Open
   (Master Alarm, SUIT/FAN Warning
   SUIT FAN Comp Lts - On)

2 CB(11) ECS: SUIT FAN 1 - Close
   H2O SEP SEL - PUSH SEP 1

3 SUIT FAN - 1 (SUIT/FAN Warning,
   FAN Comp Lts-Off,ECS Caution,
   H2O SEP Comp Lts -Off In 2 min)
   CB(16) ECS: SUIT FAN 2 - Close

355 57
298.00
-57 57

360 00
302.43
57.57

97:28

GLYCOL PUMP CHECK

1 CB(11) ECS: GLYCOL PUMP 1 - Open
   (Master Alarm, ECS Caution
   Lt - On Momentarily)
   CB(11) ECS: GLYCOL PUMP 1 - Close
   (GLYCOL Comp Lt-On)

Basic Date _____ 2/6/70
Changed _____

Basic Date _____ 2/6/70
Changed _____
ACT 31   -31

OE ZERO CDU (NO ATT Lt-Off)
CSM ATT HOLD No Longer Required

7E
7 SET REFSMFLG
000E,1E, V01 N01E,77E Confirm
13 Is Set (Set If 1st Digit Is
5, or 7)

1E

OE

O On LM MARK - ENTR
ime; Copy CSM & LM OG, IG, MG

:08:06

69 CM 163.42 CM 345.67 CM
26 LM 345.92 LM 011.79 LM

imbal Angles And Time To MSFN

2 GL
   CB

3 GL

   CB

   GL

4 Bi

[Handwritten note:] This pub. was utilized to transfer CSM guidance data to LM guidance system so the spacecraft data of our attitude with respect to the celestial sphere would not be lost. Note the time these calculations were made GET 58 08 06 about two hours after the explosion. *Jim Lovell*

VHF B CHECKOUT

1 CSM Configure for VHF Simplex B
   VHF B XMTR - VOICE
   VHF B RCVR - ON
   VHF ANT - FWD
   AUDIO (Both): VHF B - T/R

2 Both CDR & LMP Perform Voice Check
   On VHF Simplex B

OUT OF AFRICA

Evidence from fossils, ancient artefacts and genetic analyses combine to tell a compelling story of the migration of anatomically modern humans. Two possible routes have been identified for the human exodus out of Africa. A northern route would have taken our ancestors from their base in eastern sub-Saharan Africa across the Sahara desert, then through Sinai and into the Levant. An alternative southern route may have charted a path from Djibouti or Eritrea in the Horn of Africa across the Bab el-Mandeb strait and into Yemen and around the Arabian peninsula.

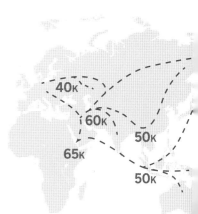

THE MATHS THAT SAVED APOLLO 13
The Lunar Module Systems Activation Checklist of Apollo 13 shows Lovell's handwritten numbers, which were part of the calculations made just two hours after a Service Module's oxygen tank explosion left them marooned in space. Lovell explains it briefly in the blue post-it note.

FLOWS
OF GENES
AROUND
THE GLOBE

- - - - - 10ᴋ
10,000
YEARS AGO

16ᴋ

15ᴋ

why the United States reaped such a colossal economic reward from its investment in Apollo. The generation of scientists and engineers who worked on and were inspired by Apollo went out into the wider economy and delivered a huge investment return; a series of studies, including one by Chase Econometrics, showed that for every dollar invested in Apollo, at least $6 or $7 was returned as increased GDP growth. This should, of course, be bloody obvious – new knowledge grows GDP – but every generation of politicians seems to require re-educating to understand the difference between spending and investment. And while I'm polemicising, let me say that the usual political argument – that public support is needed for such large investments – is drivel. Firstly, the investment in NASA wasn't that large, never exceeding 4.5 per cent of the Federal budget throughout the lifetime of Apollo. And secondly, it is a politician's job to lead from the front. Make the case that investment in knowledge, in pushing the boundaries of human capabilities and exploring all frontiers, both physical and intellectual, is the key to the future wealth, prosperity and security of civilisation. Aspire to be Kennedy, not a hand-wringing apologist for intellectual and technological decline.

The nine Apollo flights to the Moon remain the furthest modern humans have explored beyond the Rift Valley in our 200,000-year history. Homo sapiens first left Africa in large numbers 60,000 years ago, so on geological timescales we didn't hang around. Our ancestors followed waves of earlier hominins. Homo erectus were in South East Asia 1.6 million years ago, and half a million years later Neanderthals had colonised Europe and Homo floresiensis were in Southern Asia. The details of the migration 60,000 years ago are particularly well understood as a result of the combination of genetic, archaeological and linguistic studies. The precision comes in part from the tracking sequences of mitochondrial DNA, which is passed down from the mother and not shuffled by sex. This makes it relatively stable and easy to track – changes are caused by mutations alone. The most widely accepted interpretation of the data suggests that a small population of between 1000 and 2500 individuals left East Africa 60,000 years ago and moved north across the Red Sea and through Arabia. The group then split, moving into Southern Europe 43,000 years ago, and travelling through India and into Australia on roughly the same timescale. The crossing into North America, via eastern Russia, was probably later, around 15,000 years ago.

These early groups of humans were hunter-gatherers. It has been estimated that the basic social units would have reached a maximum of around 150 individuals. This is known as Dunbar's number, after the British anthropologist Robin Dunbar, who suggests that the largest social group amongst any given population of primates is related to the size of their brains (specifically the neocortex). Dunbar's number can be observed today in the size of the average person's social network, both in the real world and online; our hardware – the brain – has not changed appreciably since the first humans appeared in Africa 200,000 years ago. These social groups would have lived in loosely bound tribes, perhaps reaching a size of between one and two thousand individuals, operating within an area of around 100 kilometres. Populations would stabilise, perhaps in response to social factors, but also as a result of increased mortality rates caused by parasitic diseases and diminishing per-capita resource availability, before fragmenting and spreading. In this fashion, the rate of progression of our ancestors across the globe has been estimated to have been around 0.5 kilometres per year, or 15 kilometres per generation. Population density did not rise significantly beyond these levels until these proto-societies shifted from a hunter-gatherer lifestyle to agriculture around 12,000 years ago. This shift was the trigger for the development of civilisation: the most important single step, following the migration out of Africa, in the journey from apeman to spaceman.

HOMO ERECTUS/ERGASTER
There are differences in the literature concerning the appropriate classification of Homo erectus in Africa and in Asia. Homo erectus (sensu lato) is the term used to describe both the African and Asian populations. The population in Africa is sometimes referred to as Homo ergaster, and the Asian population as Homo erectus sensu stricto. What is widely accepted is that the African population evolved in the Rift Valley from Homo habilis around 1.9 million years ago, and subsequently spread into Asia.

FARMING: THE BEDROCK OF CIVILISATION

There are many competing theories as to the reason for the domestication of crops, but many note the correlation between the first evidence of agriculture and the beginning of the current inter-glacial period known as the Holocene, 12,000 years ago. In the fertile crescent around modern-day Jordan and Syria, people known as the Natufians were beginning to settle into larger communities, perhaps because of the relatively benign climate. The area would have been forested and rich in wild cereals, fruits and nuts, rather than the austere desert of today. One theory is that a brief 1000-year cold period known as the Younger Dryas, beginning around 10,800 BCE, triggered drier conditions in the region, forcing the Natufians to begin cultivating the previously abundant wild crops on which they had come to rely. Whatever the reason, it is generally agreed that the foundational crops of modern agriculture, including wheat, barley, peas and lentils, were all to be found in the Fertile Crescent by 9000 BCE, and by 8000 BCE the banks of the Nile were being cultivated. At approximately the same time, evidence of farming can be found in Asia's Indus Valley, in China and in Mesoamerica. This suggests that there was no single environmental or developmental cause for agriculture, because it appeared independently at many sites across the world. Rather, our large brains and relatively large social groups were ready to take up the challenge when the need arose.

Once agriculture was established, larger numbers of people could live together, taking advantage of the more stable food supply. The freedom from continual hunting and gathering would have introduced a new aspect to human life – free time – and it was used to great effect. Some of the earliest farmers settled in a place known as Beidha in modern-day Jordan around 7000 BCE. Living in round, stone-built houses, they grew barley and wheat and kept domesticated goats, engaged in ritual and ceremony and buried their dead. Importantly, each of these activities was carried out in specific areas of the settlements: the beginning of 'town planning'. By the 2nd century BCE a Semitic people known as the Nabataeans lived around Beidha. They employed new technologies to increase the reliability of farming and constructed walled agricultural terraces on the hillsides around the village to collect and store water. Animal husbandry was also expanding, with the domestication of cows, pigs, donkeys and horses. Even previously dangerous animals were coerced into living with humans; there is evidence that the Nabataeans kept dogs. As the great empires of Egypt, Greece and Rome prospered, the Nabataeans remained partially nomadic, driving their camel trains across the desert along the long-established trade routes between North Africa and India and the great cities of the Mediterranean. But then, around 150 BCE, they decided to try something different. A few kilometres south of Beidha, in a narrow gorge naturally formed in the soft sandstone rock, they built the city of Petra.

Today tourists stream through a magnificent passageway lined with buildings carved out of the desert rocks and known as the Siq, but 2000 years ago the great and good of Mesopotamia, Rome and Egypt would have walked this route into this jewel of late antiquity.

The grandeur of the buildings is still overwhelming; they stand not as great architecture for their time, but as simply great, with no caveat. The most famous is called Al Khazneh, which means 'Treasure Box', because of the carved urn above the entrance which, Bedouin legend has it, contains the treasure of a Pharaoh. Monumental architecture is a common feature in the rise of human civilisation. It is a statement of power and grandeur to impress and cow outsiders, but it also serves an internal purpose, cementing the position of the rulers in the hierarchy and therefore providing the stability and security on which civilisation rests.

LOST CITY
Petra was the capital of the Nabataean kingdom and was carved into a red sandstone gorge over 2000 years ago in what is now Jordan. The hillsides running down the valley from the carved tombs are scattered with rocks, but closer inspection reveals them to be bricks, the remains of houses, temples and palaces.

At its peak, Petra had a population of thirty thousand. Today, it lies empty and abandoned, as it has for nearly 1500 years. Its only occupants are a handful of Bedouin tribespeople who have made their homes amongst the ruins.

Over time, a virtuous circle emerges; the buildings help the civilisation prosper, and the more prosperous the civilisation, the more impressive the buildings become.

Petra's wealth was derived from its location. Built within a natural gorge, the area is prone to flash floods, which provided precious water in a landscape that was arid by the time the Nabataeans began to build. The city also sits at the fulcrum of the ancient nomadic trade routes along which wood, spices, incense and dyes were transported from Africa and India and into the great Mediterranean civilisation beyond. The appetite of the Greeks and Romans for exotic goods was insatiable; black pepper alone fetched forty times its own weight in gold in a Roman market. Petra, because of its strategic location, controlled all that trade and taxed it. Today, 1500 years after the city was abandoned, it is still a magnificent site – an overused but entirely accurate statement. Talk to an archaeologist, however, and you quickly realise how much more impressive it would have been in its heyday. The hillsides running down the valley from the carved tombs are scattered with rocks, but closer inspection reveals them to be bricks, the remains of houses, temples and palaces. Everything from Al Khazneh to the houses would have been covered in white plaster and painted in bright colours which would have appeared resplendent against the monochrome desert sands.

To build on this scale required a huge labour force; Petra was home to at least thirty thousand people living in a few square kilometres of desert. Such a population density required technological innovation on a metropolitan scale, and the Nabataeans, perhaps more than any other civilisation in antiquity, were masters of fluid engineering. Virtually every drop of rainwater that fell on the surrounding hillsides was captured in grooves and stored in giant reservoirs and cisterns. They were better at plumbing than the Romans, who employed the Petran engineers in Rome. Petra had the world's first pressurised water system, which could deliver 12 million gallons of water a day into the city.

Outside the city, the irrigation system continued out into the surrounding fields, lining the hillsides in still-visible terraces; the Nabataeans didn't simply build a city, they terra-formed a landscape. I stood and imagined the ancient valley views with some awe; the mountain slopes would have been green with maize, barley, pulses and vineyards – a desert turned green and feeding this grandest of desert civilisations for six centuries. Whenever I see the ruins of Petra, Rome, Athens or Cairo, I wonder what Earth would be like today if the great civilisations of antiquity had not fallen. I blame the philosophers for not discovering the scientific method earlier and inventing the electric motor. How hard can it be?

Agriculture, then, was fundamentally important to the rise of civilisation because it enabled large numbers of people to live in one place, and gave them access to resources and time, which would have been unavailable to hunter-gatherers. With resources and time comes the division of labour, freeing up a small but important subset of individuals to engage in pursuits other than those necessary for immediate survival. Farmers, stonemasons, priests, soldiers, administrators and artisans emerge, together with a ruling class who begin to direct the construction of monumental architecture, partly for their own selfish ends. And cities like Petra become possible.

Petra was a relative latecomer in the emergence of the cities and civilisations of antiquity. The first great ancient civilisation, the Old Kingdom of Egypt, arose around 2600 BCE along the fertile and farmed banks of the Nile, and precisely the same pattern of agriculture, followed by social stratification, ritual and monumental architecture, can be seen. Present also in the Old Kingdom, and possibly developed there, was the one final vitally important innovation we will soon discuss: the written word.

TOWERING STRUCTURES
The Monastery (Ad Deir) at Petra dates from the 1st century BCE and is the ancient city's largest monument. Architecturally it is an example of the Nabataean Classical style. It may have been used as a church or monastery by later societies but most likely began as a temple.

THE KAZAK ADVENTURE: PART 1

It all seemed so simple when written down on a piece of paper. The BBC prepares something known as a call sheet, which tells a film crew everything they need to know about a trip. Call sheets are very neat; all the timings work beautifully, carefully documenting flights, ground transfers to locations and filming and rest periods, all of course in accordance with health and safety regulations and all that. Things never quite work out the way they're envisaged back in the office, of course, but filming the return of the Expedition 38 crew from the International Space Station to the Kazak Steppe in March 2014 was the wildest adventure I've experienced.

The call sheet said that we would fly into Astana on 8 March, arriving at 1am on the 9th into our hotel. After a leisurely breakfast at 9am, we'd drive to a city called Karaganda, which has a spectacular statue of Yuri Gagarin in the town square. There, we'd meet up with our drivers who, embedded with Roscosmos, the Russian space agency, would drive us out to the landing site the following morning, arriving in time for a 'hot meal' and a good rest on the Steppe, ready to film the landing on the morning of the 11th after, of course, a 'hot breakfast'. We'd then drive back to the airport, hop on a flight, and be home in time for lunch on the 12th. A doddle. Bollocks.

The Steppe of central Kazakhstan in March is a featureless frozen wilderness covering around 800,000 square kilometres of the country's interior. There are no towns and few roads; just tufts of stunted brown grass and snow fading into an ice-grey leaden sky. In March 2014 temperatures were unseasonably cold, falling below -20°C at night, and it was snowing. Our team had standard 4×4 vehicles, which got stuck in the snow by mid-afternoon the day before the landing, even though we'd set off three hours earlier than the 6am officially sanctioned health and safety call time because of the weather. This was problematic, because our *Apeman Spaceman* film was constructed carefully around this moment – the return of three human beings from space. Over vodka, cold meat and bread, we discussed our options.

We'd been helped along the snowy roads by a Russian team from the Siberian city of Tobolsk in two spectacular 6-wheel-drive vehicles, hand-built by a company called Petrovich. Tobolsk is best known for being the place dissidents were sent during the Soviet era. Tsar Nicholas II and his family enjoyed the Tobolskian hospitality for a year before being transported to Ekaterinburg to be shot. Mendeleev, the inventor of the Periodic Table, was born there, but so was Rasputin. It's a tough place, and they know how to build tough vehicles. Our guide from Roscosmos managed to radio the Petrovich team, and they agreed that if we could catch them up in the frozen wilderness, they could take two of us out to the landing site. The cameraman and I jumped aboard a pair of snowmobiles, and headed out into the rapidly dimming late-afternoon twilight in search of the men from Siberia. If we hadn't found them, then presumably you wouldn't be reading this, but we did.

It was a difficult decision to jump onto the snowmobiles. We didn't have a satellite phone because they are illegal in Kazakhstan, and nobody spoke English so we couldn't quite assess the level of difficulty associated with finding these two Siberian needles in a Kazak Steppe. And we didn't know who the Siberians actually were. It seemed that they were free-lancers, hired to take photographs and broadcast live television pictures back from the landing site for the Russian space agency. We also had to decide whether we could make the film with only two people. Much as I spend a lot of time dreaming about jettisoning directors, producers and executives, there is a reason why we usually take a crew of six. Sound is particularly important; you don't really miss the soundman until he's not there (our soundman on the series is called Andy, but we always called him soundman – there are too many other things to remember).

As it turned out, the Petrovich crew were a hospitable and professional bunch, although their willingness to spend many days out in the wilderness waiting for the Soyuz – they'd driven down from Siberia and were in no hurry to get home – played on our minds. Approaching midnight on the night before the landing, we received a message from Roscosmos that the landing might be postponed due to the poor weather, and the decision was made to camp out on the Steppe and wait. In the distance, we could make out a small group of farm buildings through the snow, and we headed towards them. In broken English, one of the crew told us that it is a Kazak tradition to welcome travellers into your home, at any time of the day or night, and offer them food. And so we found ourselves inside a farm house that appeared to have heated walls and resembled the inside of an oven, eating a feast of jam, bread, assorted sweets and horse, all washed down with vodka, which the Petrovich crew carried in large crates alongside their satellite broadcasting hardware. It was unforgettable. *Human Universe* was filmed as a love letter to the human race, and time and again when I've found myself immersed unexpectedly in a culture, I've been reminded about why it is appropriate to want to write one.

At 4am, soaked in vodka, the call came through. Commander Oleg Kotov, Sergey Ryazansky and Mike Hopkins had climbed aboard the

BATTLING WITH
THE ELEMENTS
Filming the return of the
Expedition 38 crew from the
International Space Station to
the Kazak Steppe was one of
the wildest adventures I have
ever experienced. That we got
to see it is a testament to the
determination of man!

Soyuz and were preparing to depart the International Space Station. I was elated, because I genuinely thought the landing would be called off, and I had no idea what that would have meant, other than waiting for the storms to clear on the Steppe.

At 6.02am Kazak time, the Soyuz TMA-10M, the 199th Soyuz to fly since 1967, undocked from the ISS. This is the point of no return, except in an emergency. Just 2 hours and 28 minutes later, it fired its engine for a pre-programmed burn of 4 minutes and 44 seconds. This reduced the spacecraft's velocity by 128 m/s relative to the Station, which in its orbit that day was travelling at 7358 m/s. That number is not arbitrary. It is given by a simple equation which can be derived easily from Newton's Law of Gravitation and his Second Law of Motion, $F=ma$. We leave it as an exercise for the reader to show that these two laws of nature can be re-arranged to show that the velocity v of any object in a circular orbit a distance r from the centre of the Earth, mass M_e, is given by

$$v = \sqrt{\frac{GM_e}{r}}$$

To derive this result, you need to know that the force required to maintain an object of mass m in a circular orbit is mv^2/r.

BACK FROM SPACE
A Soyuz spacecraft returning from the ISS is shown as it safely lands in a remote part of the Kazak Steppe.

SAFE LANDING
A search and recovery team helps to extract ISS crew members from the spacecraft. Re-entry is an extremely physical experience, and Earth's gravity a powerful returning force.

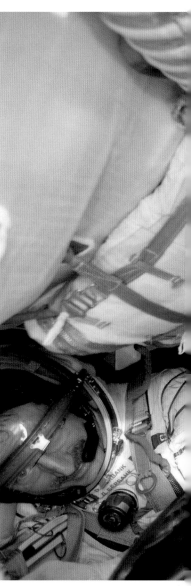

The Space Station orbits at an altitude of between 330 and 445 kilometres – let's choose the middle ground of 387 kilometres – this is a back-of-the-envelope calculation. 'Estimate is the name of the game', as my old physics teacher used to say at school. The radius of the Earth is 6,378 kilometres, and the mass of the Earth is 5.97219×10^{24} kg. Newton's gravitational constant is 6.67384×10^{-11} m^3 kg^{-1} s^{-2}. Do the calculation yourself; maths is good for you. With these numbers, v is approximately 7675 m/s, which is close enough – the difference is due to the precise altitude of the ISS that day. I love doing little calculations like this. They reveal the immense power of mathematical physics; this really is the orbital velocity of the International Space Station, and it is forced to be so by laws of nature first published by Isaac Newton in 1687. If you've never done a calculation like this before, you should feel elated. The biologist Edward O. Wilson called this feeling the Ionian Enchantment, a poetic term he introduced to describe the realisation, credited to Thales of Miletus in 600 BCE, that the natural world is orderly and simple, and can be described with great economy by a small set of laws. It is nothing short of wonderful that we can calculate the orbital velocity of the International Space Station together in a few lines of a popular book, and this points us neatly towards the story of the last great innovations in the ascent from apeman to spaceman: the written word.

CLOSE QUARTERS
ISS crew members, US astronaut Daniel Burbank and Russian cosmonauts Anton Shkaplerov and Anatoly Ivanishin are seen inside the Soyuz capsule on another mission in 2012, shortly after landing in Kazakhstan.

INTERMISSION: BEYOND MEMORY

WRITING HISTORY
The earliest known system of writing is generally accepted to be cuneiform, the Sumerian system that emerged around 5000 years ago in Mesopotamia. Here cuneiform inscriptions are seen on a stairway wall in the Palace of Darius I, at the ancient Achaemenid ceremonial centre of Persepolis, in Iran.

I began my degree at the University of Manchester in 1992, which is when I started doing physics full time. I gained my PhD in 1998, spent the next 11 years working as a particle physicist at the DESY laboratory in Hamburg, Fermilab in Chicago and CERN in Geneva. In 2009 I began filming *Wonders of the Solar System*, which slowed down my research a bit. But I've been at it now for 22 years, which is almost half my life. In that time, I've learnt a lot about how to be a scientist, how to think about scientific problems, how to make measurements of nature, particularly the behaviour of subatomic particles, and how to interpret those measurements to generate new knowledge and make new discoveries. But given all that, there is no way that I would be able to calculate the orbital velocity of the International Space Station from scratch. Given Newton's laws, it's trivial. Without them, it would be virtually impossible. Newton's laws are far from obvious; they took Newton a scientific lifetime to produce, and he was a genius – one of the greatest scientific minds of all time. And even he didn't start from scratch. He relied heavily on the previous works of Galileo, Euclid and a hundred other philosophers, geometers and mathematicians whose names have been forgotten but whose works remain as cornerstones of our scientific culture. The reason we could run through that simple calculation together is that the thoughts and discoveries of these generations of philosophers, scientists and mathematicians were not lost; they were preserved forever in the written word.

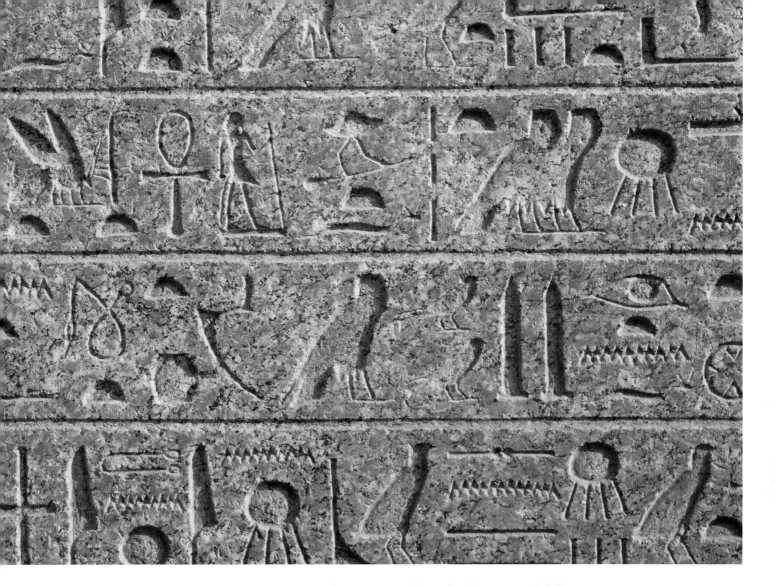

Writing appears to have arisen independently in several different cultures, just as with the development of agriculture, and just as agriculture triggered the birth of civilisation 12,000 years ago, so the emergence of writing supported a rapid increase in the complexity of civilisation. The earliest known system of writing is generally accepted to be Cuneiform, the Sumerian system that emerged around 5000 years ago in the cities of Mesopotamia, although it is possible that Egyptian hieroglyphs may predate it. Literally meaning 'wedge-shaped', cuneiform comprises a thousand or more symbols created using a stylus made from reed that was pressed into a soft clay tablet. Following cuneiform and hieroglyphs, other forms of script emerged in Greece, China, India and, later, Central America.

Writing seems not to have arisen out of a deep human need to share and record intimate thoughts and lay down knowledge for future generations; that would be far too romantic. Rather, it appears to have served a more practical purpose, revealed in a set of around 150 Nabataean scrolls discovered by archaeologists in 1993. The scrolls date from around 550 CE, in the final period before Petra was abandoned. One of the most intact documents relates to a court case between two priests. It is alleged that one of the priests decided to run away from their shared house, taking a key to one of the upstairs rooms, two wooden beams that presumably held the roof up, six birds and a table. This is probably how writing began; the invention upon which modern human history rests arose, disappointingly, for admin purposes. This is seen not only in the relatively late Nabataean scrolls, but in many of the early texts. Cuneiform

HIEROGLYPHIC WRITING
Hieroglyphs ('sacred carving') are symbols, here inscribed in rock, which made up ancient Egyptian writing. This carving features the profiles of many different types of bird, as seen at the Temple of Karnak in Luxor, Egypt.

developed because of a need to keep track of trade and accounts in the increasingly complex economy of Mesopotamia. Egyptian hieroglyphs may be an exception, as there is a strong ritual component, but there is also evidence of their early use in commerce, administration, trade and law – the foundations of a modern society. Information about the natural world was also recorded; in hieroglyphs we see the cycle of the seasons chronicled, as well as important environmental events. There are also some beautiful early examples of the use of writing to express deeper human desires and feelings that resonate strongly today and show, yet again, that our ancestors had inner lives not too distant to our own.

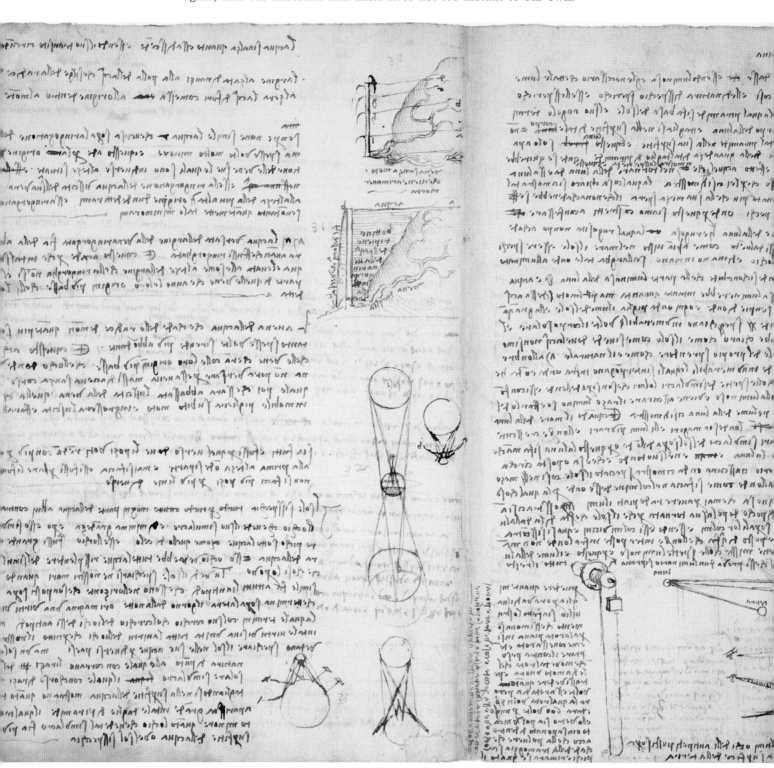

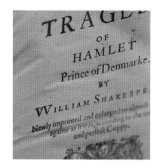

But the oldest surviving papyrus documents from the Old Kingdom are marvellously prosaic. From Dynasty 5, in the reign of Pharaoh Djedkare-Izezi between 2437 and 2393 BCE, can be found an early version of the parrot sketch.

'As Re, Hathor and all the gods desire that King Izezi should live forever, and ever, I am lodging a complaint through the commissioners concerning a case of collecting a transport-fare.'

And so the letters continue; whilst the tombs are covered in the names of pharaohs and stories of the Gods, the people of Egypt were using writing as we do today, and I find it wonderful, reassuring and moving in a funny sort of way to hear ancient voices complaining down the years. Perhaps we humans really will never change. From Dynasty 20, a millennium later during the reigns of Ramesses III and IV between 1182 and 1145 BCE, the complaints continue.

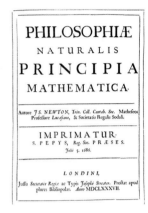

'The scribe Amennakht, your husband, took a coffin from me saying, "I shall give the calf in exchange for it", but he hasn't given it until this day. I mentioned this to Paakhet, who replied "Give me a bed in addition to it, and I will bring you the calf when it is mature". And I gave him the bed. Neither the coffin nor the bed is yet here to this day. If you are going to give the ox, send it on; but if there is no ox, return the bed and the coffin.'

Alongside the letters, the ritual, the complaints, the admin and the legal documents, there was also a sophisticated literary and storytelling tradition in Ancient Egypt, and a powerful appreciation of the value of the written word. Three thousand years ago on the banks of the Nile, during the reign of Queen Twosret, someone wrote a Eulogy for the writers;

These sage scribes ...
Their names endure for eternity,
Although they are gone, although they have completed their lifetimes, and all their people are forgotten.
They did not make for themselves pyramids of bronze with stelae of iron ...
They made heirs for themselves
as the writings and Teachings that they begat ...
Departing life has made their names forgotten;
Writings alone make them remembered.

Taken from The Tale of Sinuhe and other Egyptian Poems
1940–1640 BC, *Oxford World's Classics*

Writing was the final pivotal moment in our ascent from early agrarian civilisations to the International Space Station, because it frees the acquisition of knowledge from the limits of human memory. The hardware restrictions set down in the Rift Valley 200,000 years ago no longer matter. Writing allows a practically unlimited amount of information to be passed from generation to generation, and to be shared across the world. Knowledge is no longer lost but is always added to; it becomes widespread, accessible and permanent. A little boy from Oldham, Lancashire, can inhabit the mind of Newton, assimilate his lifetime's work and derive new knowledge from it. Writing created a cultural ratchet, an exponentiation of the known that allowed humanity to innovate and invent way beyond the constraints of a single human brain. We now work together as a single mind spread across the planet and with a memory as long as history. It is this collective effort, enabled by the written word, that carried us, the human race, paragon of animals, from the Great Rift Valley to the stars. I deliberately borrow from Shakespeare; the most precious objects on Earth are not gems or jewels, but ink marks on paper. No single human brain could conceive of *Hamlet, Principia Mathematica* or *Codex Leicester*; they were created by and belong to the entire human race, and the library of wonders continues to grow.

SHARED WISDOM
The written word has enabled great minds such as Leonardo da Vinci, William Shakespeare and Isaac Newton to pass on their knowledge, vision and findings to future generations, to be shared across the world.

THE KAZAK ADVENTURE: PART 2

The drive from the farm to the Soyuz landing site was agricultural. The Petrovich vehicles work as a pair, dragging each other out of snowdrifts when they get stuck. I wondered through the mildly paranoid haze that descends after 48 hours of wakefulness and 48 shots of vodka (which is not optional if Russian sensibilities are to be respected) what would happen to us if both vehicles got stuck. By dawn, we arrived at the GPS coordinates given to us by Roscosmos, and waited. We knew precise timings for re-entry, because those are given by physics alone once the de-orbit burn of 4 minutes and 44 seconds occurs. Recall that the Soyuz, along with the Space Station, was in a circular orbit travelling at 7358 m/s, and the engine burn slowed it down by precisely 128 m/s. This put the Soyuz into an elliptical orbit, which, when the breaking effects of the atmosphere are taken into account, put the craft on a collision course with Kazakhstan. It's quite simple, and it works. In my experience filming with Roscosmos, the words 'it's quite simple and it works' sum up Russia's successful half a century in space. They don't do things in as shiny, hi-tech a fashion as the United States; the Soyuz has been flying astronauts into space with minimal design changes since 1967. But today, the Soyuz is the only way to get to and from the ISS, and it is a reliable system. But to my inexperienced eyes, unused to the way the Russians do things, the return of the Expedition 38 crew after six months in space felt like a traction engine rally in Yorkshire arranged by Fred Dibnah. That's not meant as a criticism, because I'd trust Fred Dibnah to organise a traction engine rally, and I'd trust the Russians to get me back from space. But neither stands on ceremony.

At precisely 9.23am, the Soyuz emerged from the snow-filled skies above the Steppe, swinging from its parachutes, and touched down with a burst of soft landing jets. One of our Petrovich colleagues saw it with his binoculars, and we headed off towards the spaceship in the snow. In one of the most bizarre moments of my life, we arrived, and, without thinking, jumped out and stumbled through the drifts towards the spacecraft. I fumbled around with the microphone for a while (recall that soundman didn't make it), and then realised that there were no other vehicles around. A single helicopter had just landed; apart from that, there was only the wind driving gentle flurries across the Steppe.

Minutes later, the support vehicles arrived and Oleg Kotov, Sergey Ryazansky and Mike Hopkins were dragged from the hatch of their Soyuz, wrapped in sleeping bags and put into deckchairs. They looked happy, but knackered, and mildly discombobulated as a parade of Russian army generals in very big hats seized the opportunity for a photo. The Russians don't overdo things; they just do them. Five times a year men and women make this voyage back to Earth having spent half a year in space, living amongst the stars on the International Space Station. Since the first expedition began on 2 November 2000, the station has been continuously occupied, and I hope that there will never again come a time when every human being is confined to Earth.

I carried in my pocket a reminder of my time in Ethiopia, the small flint we used for filming in the Rift Valley. I imagined a human, my great, great-grandfather, sitting somewhere in the vicinity of what would one day become Addis Ababa, diligently chipping away at the obsidian in my hand, the whole of history away. I set it down in the snow next to the Soyuz, descended from it as I am from him.

SPACEMAN
When I was young I dreamt of being an astronaut – it was why I became interested in astronomy and physics. Floating in a most peculiar way in the ISS simulator tank is the closest I will ever get.

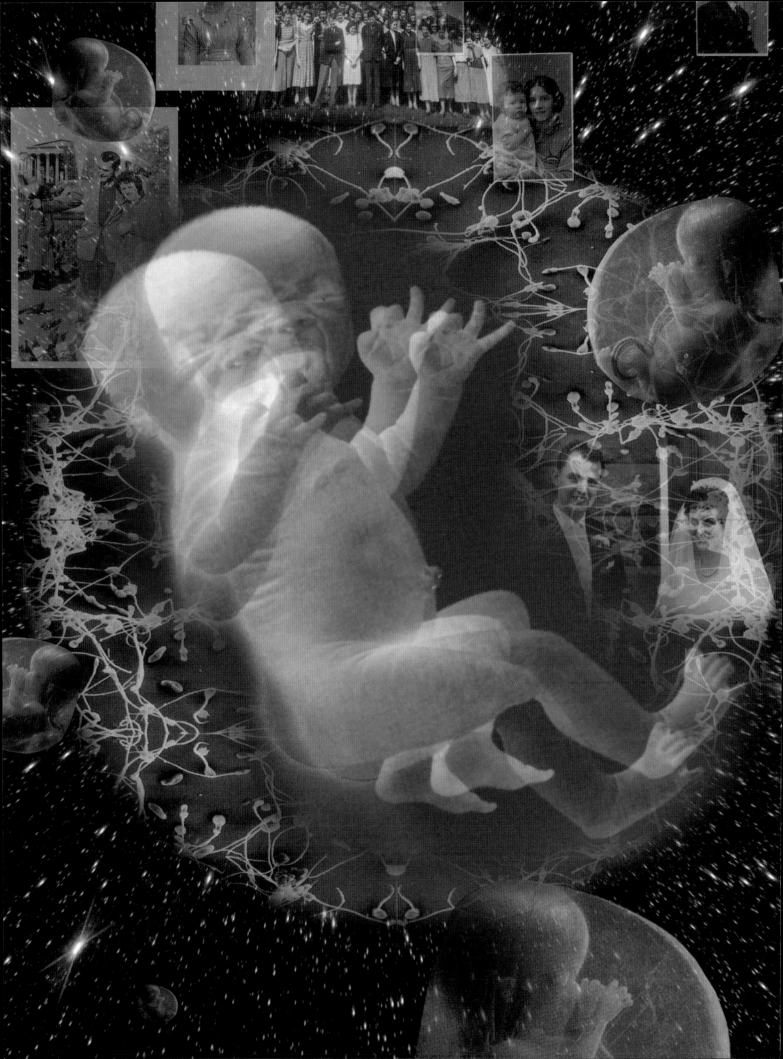

WHY ARE WE HERE?

But, after all, who knows, and who can say
Whence it all came, and how creation happened?
The gods themselves are later than creation,
so who knows truly whence it has arisen?

Ancient Brahmin Verse

A NEAT PIECE OF LOGIC

There is tension at the interface between science and language. Language is concerned with human experience. Everyone understands what is meant by questions such as 'Why are you late?' 'I'm late because my alarm clock didn't go off'. But this answer is incomplete, and could be followed by a series of further questions in an attempt to establish precisely why.

'Why didn't it go off?'

'Because it's broken.'

'Why is it broken?'

'Because a piece of solder melted on the circuit board.'

'Why did the solder melt?'

'Because it got hot.'

'Why did it get hot?'

'Because it's August and my room is hot.'

'Why is it hot in August?'

'Because of the details of the Earth's orbit around the Sun.'

'Why does the Earth orbit the Sun?'

'Because of the action of the gravitational force.'

'Why is there a gravitational force?'

'I don't know.'

All scientific 'Why?' questions end with 'I don't know' if you keep pushing far enough, because our scientific understanding of the universe is not complete. The most fundamental description we have for anything comes down to a set of theories describing the smallest known building blocks of the universe and the forces of nature that allow them to interact with each other. These theories are known as laws of physics, and when we ask about the origin of these laws, the answer is 'We don't know'. This is because in the Big Bang model, our understanding of physics before 10^{-43} seconds after the origin of the visible universe is virtually non-existent, and the origin of the laws lies at some point before that. 'The laws themselves are later than creation, and who knows truly whence it has arisen'. Our best theory of space and time, Einstein's General Theory of Relativity, no longer applies at the earliest times; the conditions were so extreme in those first moments, known as the Planck epoch, that some kind of quantum theory of gravity, which we do not possess, will be needed to describe it.

The universe is now 13.798 +/- 0.037 billion years old, according to our current best measurements and theoretical understanding, and has been gently expanding and cooling ever since the Big Bang. The universe appears to be gently increasing its expansion rate, and approximately 68 per cent of the energy in the universe is associated with this sedate acceleration. The energy has a name – dark energy – but its nature remains one of the great unsolved challenges for twenty-first-century theoretical physics. Of the 32 per cent that remains, approximately 27 per cent is in a form of matter known as dark matter. The nature of this is also unknown, but it probably comes in the form of as yet undiscovered subatomic particles. The remaining 5 per cent makes up the stars, planets and galaxies we see in the night sky, and of course human beings. The part of the universe we can see is around 93 billion light years across and has reached a relatively chilly temperature of 2.72548 +/- 0.00057 Kelvin due to its expansion.

The question of the origin of the universe is an old one in philosophy, and often framed in terms of the 'First Cause' argument. Leibniz is associated with a 'proof' of the existence of God in this context, which goes something like this:

Everything that exists must have either an external cause or must be eternal. If there are eternal things, then they must necessarily exist,

EXPLORING THE BIG BANG
In March 2014 this telescope, the BICEP2, detected a pattern in the cosmic microwave background which supports the inflation theory attributed to the Big Bang, the origins of our universe which we are still exploring.

because they don't have a cause. Since the universe exists and is not eternal, it must have an external cause, and to avoid infinite regress that cause must be an eternal and necessary thing, which we'll call God.

This is quite a neat piece of logic, obviously, because Leibniz wasn't an idiot. I don't consider such questions to fall necessarily within the domain of science. Rather science is concerned with answering more modest questions, and this is the reason for its power and success. The goal of science is to explain the observed features of the natural world.

יְהֹוָה:

Et vidit Deus lucem quod esset bona.

Mundus Intellectualis

SYLVA SYLVARVM
or
A NATVRALL HISTORY
In ten Centuries.
Written by the right Hon.ble Francis
Lo: Verulam Viscount S.ct Alban.
Published after y.e Autho.rs Death
by W: RAWLEY D.or of Diui:
nity. &c

Tho: Cecill sculp:

Anno

LONDON
Printed for W. Lee and are to be sould at
the Great Turks head, next to the Mytre
Taurne in Fleetstreet

1627

By 'explain', I mean 'build theories that make predictions that are in accord with observation'. This is a humble idea; there is no *a priori* aim to discover the reason for the existence of our universe or to build theories of everything. Science proceeds in tiny steps, attempting to find explanations for the blue sky, the green leaves of plants or the stretched, red-shifted light from distant galaxies. Sometimes, those tiny steps build up to something rather grand, like a measurement of the age of the observable universe, but that's not what anyone set out to do. This is why science is more successful than any other form of human thought when applied to questions within its domain, which is the explanation of the natural world. It starts small and works its way slowly and methodically forwards, deepening our understanding in careful increments.

Our chapter title 'Why are we here?' might therefore appear to be unanswerable by science; it's too grand a question. But that may no longer be the case, because the careful steps are taking science into this territory and the scientific language is now in place to at least address the question 'What happened before the Big Bang?' This is clearly a prerequisite for being able to make any meaningful attempt to address the reasons for our existence, although it is surely not sufficient. Immediately, I have to explain a semantic distinction before a thousand philosophers throw their togas aside and prepare to engage in a naked yet civilised and eloquent battle of ideas. I am defining the term Big Bang as the astronomer Fred Hoyle originally introduced it into physics in 1949. It is to be understood as the beginning of the hot, dense state in which our observable universe once existed. Conventional cosmological theory, as described in Chapter 1, traces the evolution of the universe backwards in time, with conditions getting hotter and hotter and denser and denser until the point where we are unsure of the correct rules of physics. Currently this is earlier than approximately 10^{-10} seconds, which is associated with the current power of the Large Hadron Collider. If the universe existed in some other form before the hot, dense state came into existence 13.798 billion years ago, then that's what I'm referring to as the time before the Big Bang. Science might accidentally wander into Leibniz's territory if, for example, this time before the Big Bang were discovered to be infinite, or that the state before the Big Bang was logically necessary and describable by current or yet-to-be discovered laws of physics. Such a theory would also have to explain precisely all the properties of the universe we see today. From a scientific perspective of course, we don't care about Leibniz; it is not the role of science to prove or disprove the existence of God. Rather we are only interested in taking our careful steps backwards in time as far as the evidence and theoretical understanding allow. The exciting thing is that developments in cosmology since the 1980s now point quite firmly towards the existence of a state before the Big Bang as defined above, and that is primarily what this chapter is about.

This chapter is also about you. I suspect most of us have mused about the question 'Why are we here?' For some, the question and answer may be absolutely central to their lives. For others, myself included, it's something I used to think about on a hillside desolate beside a punctured bicycle whilst wearing a secondhand overcoat I bought from Affleck's Palace, but my existentialism faded with my hair.

Having said that, a little existentialism, like the Manchester rain, never did anyone any harm, so let's place ourselves at the centre of things for a while and explore the immense contingency of our personal existence as a warm-up for the much deeper problem of the origin of the universe itself. It's a pretty deep chapter this, so put on *Unknown Pleasures*, grab a bottle of cheap cider and let's get going.

市市（株）アットハウス代表取締役社長　加藤優次

都台東区浅草橋二ノ五ノ六　テージー株式会社　玉越進

札幌市北区屯田六条十一丁目三ノ一〇　井家律子

東京都世田谷区砧二ノ五ノ三　藤田照雄

株式会社ミズモリ　代表取締役　毎日刀志　裕和産業(株)

川崎市多摩区

東京都世田谷区

東京都品川区東大井　恩田伊都子　久保佳子

NEW DAWN FADES

It was me, waiting for me,
Hoping for something more,
Me, seeing me this time,
Hoping for something else.
Ian Curtis, New Dawn Fades, Unknown Pleasures

If, in a moment of solipsism, you decide to work out the odds of your own existence, you might come to the conclusion that you are astonishingly special. You began as a particular egg inside your mother, fertilised by a particular sperm from your father. There were 180 million sperm around that day, each with a different genetic code, only one of which became 'you' in combination with one of your mother's million or so genetically unique eggs. So without going any further, you might feel lucky. If you chose to carry on, you might factor in the odds of your parents having sex on that particular day, because sperm are constantly manufactured. Then there are the odds of them meeting at all, and the odds of them being THEM. And whilst we're picking up increasing armfuls of odds at the 1-in-a-100-million level, recall from Chapter 1 that there exists an unbroken line of your ancestors stretching back over 3.8 billion years to LUCA – the Last Universal Common Ancestor. If any one of those living things had died before it reproduced, you wouldn't exist. That's pretty lucky, but also completely devoid of any meaning at all. Yes, the odds of YOU existing are almost, but not quite, zero. But given the existence of the human race and a mechanism for procreation, someone has to be born. So whilst the probability of any given individual existing is tiny, it is inevitable that new babies will be born every day. Seen in this light, you are not special and your existence in the grand scheme of things is entirely understandable. Time for Joy Division and cider.

UNKNOWN PLEASURES
Ian Curtis from Joy Division muses on our hope for something more; the reality is that we can always strive to be more, but the laws of nature remind us that we are not special and our existence in the grand scheme of things is understandable!

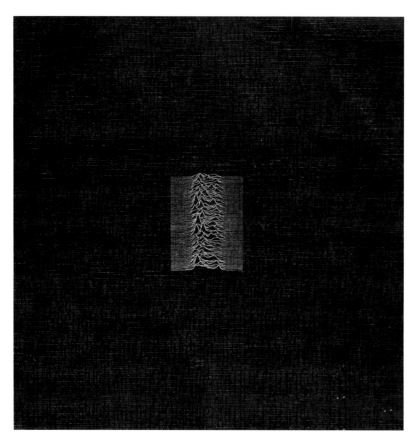

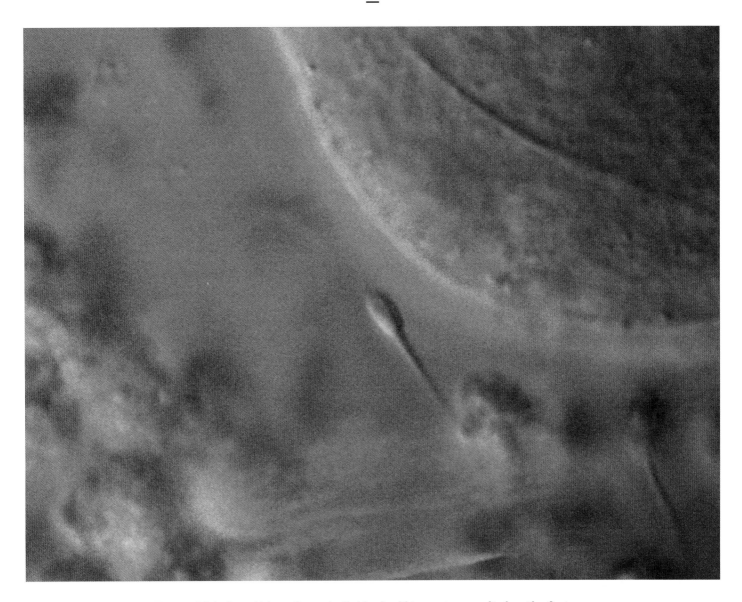

A CHANCE ENCOUNTER
Just one sperm out of 180 million will find its way into the ovum in order to produce a new human; a chance encounter which seems to defy the odds and is miraculously repeated all over the world and throughout history.

This demolition of your individual self-importance relied on the fact that a mechanism exists for the inevitable production of large numbers of human beings, given the important precondition that humans already exist. We've explored the road to human existence at length in the book already, and argued that complex multicellular life and intelligence at or beyond the level of humans may be rare in our universe. It is also clear that there are fundamental properties of the universe itself that are necessary for the existence of any form of life. The universe must live long enough and have the right properties for stars to form, and those stars must be capable of producing the chemical elements out of which living things are made, carbon being the most important. What do we mean by 'properties'? We are back to the nature of the laws of physics once again, because they describe the behaviour of matter and forces at the most fundamental level. The laws restrict the possible physical structures that are allowed to appear in the universe, and stars, planets and human beings are all examples of such possible physical structures. Questions now naturally arise; more modest perhaps than our grand 'Why are we here?' puzzle, but more amenable to scientific enquiry. How do the laws of nature allow for human beings to exist, and by how much could those laws vary before life could no longer exist in the universe?

Let us begin in the spirit of taking small steps with a brief summary of the known fundamental laws of nature.

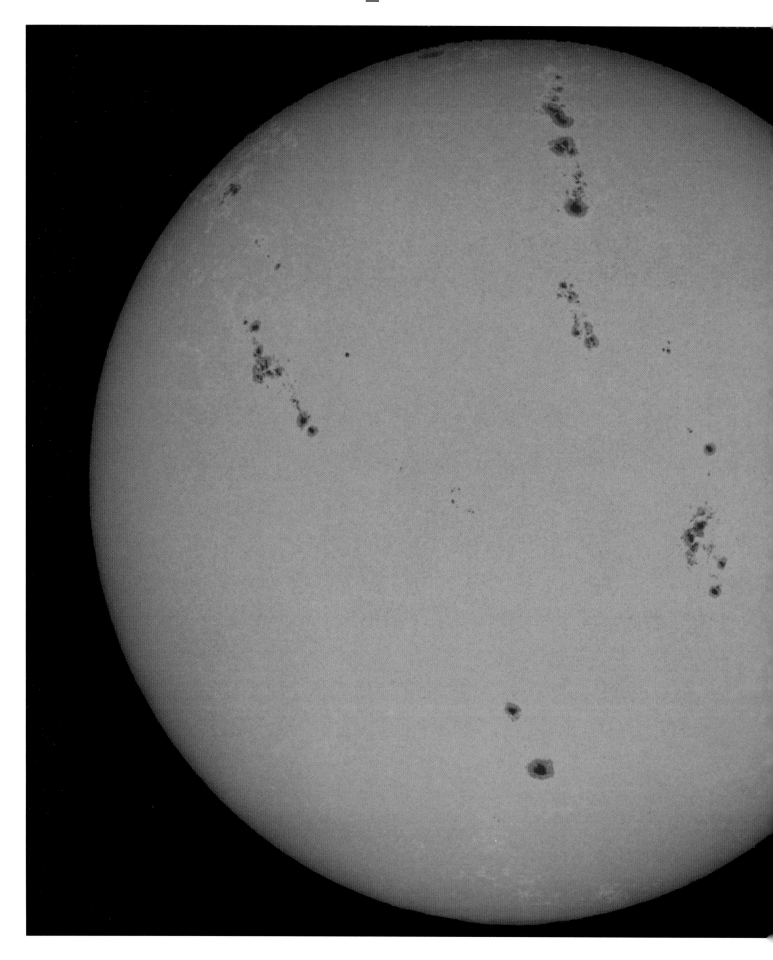

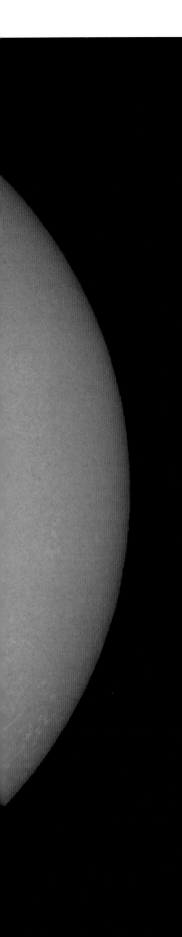

THE RULES OF THE GAME

*What really interests me is whether God
had any choice in the creation of the world.*
Albert Einstein

Attempting to describe the laws that govern the existence of everything from galaxies to human beings in a single paragraph of a book of a TV series might seem overly ambitious. It is at one level; otherwise everyone would complete physics, chemistry and biology degree courses in an afternoon. What we can do, however, is to outline the known fundamental laws in a concise and accurate way, so let us do that.

There are twelve known particles of matter, listed on page 182. They are arranged into three families, or generations. You are made out of particles in the first generation alone. Up quarks and down quarks bind together to make protons and neutrons, which in turn bind together to form your atomic nuclei. Your atoms are composed of electrons bound to those nuclei. Molecules, such as water and DNA, are built up out of collections of atoms bound together. That's all there is to you; three fundamental particles arranged into patterns. Particles called gauge bosons carry the forces of nature. There are four known fundamental forces; the strong and weak nuclear forces, electromagnetism and gravity. Gravity is missing from the top figure on page 182, and we'll get to that in a moment. The other three forces are represented in the fourth column. To see how this all works, let's focus on the familiar electromagnetic force. Imagine an electron bound to the atomic nucleus of one of your atoms. How does that binding happen? The most fundamental description we have is that the electron can emit a photon, which you can think of as a particle of light. That photon can be absorbed by one of the quarks inside the nucleus, and this emission and absorption acts to assert a force between the electron and the quark. There is a vast number of ways in which the electrons and the quarks inside the nucleus can emit and absorb photons, and these all combine to keep the electron firmly glued to the nucleus. A similar picture can be applied to the quarks themselves. They also interact via the strong nuclear force by emitting and absorbing force-carrying particles called gluons. The strong nuclear force is the strongest known force (the clue is in the name) and binds the quarks together very tightly indeed. This is why the nucleus is significantly smaller and denser than the atom. Only quarks and gluons feel the strong nuclear force. Finally, there is the weak nuclear force. This is mediated by the exchange of the W and Z bosons. All known particles of matter feel the weak nuclear force but it is extremely weak relative to the other two, which is why its action is unfamiliar, but not unimportant. The Sun would not shine without the weak nuclear force, which allows protons to convert into neutrons, or more precisely up quarks into down quarks, which has the same result. This is the first step in the nuclear burning of hydrogen into helium, the source of the Sun's energy. During the conversion of a proton into a neutron, an anti-electron neutrino is produced along with an electron. The neutrino is the remaining particle in the first generation we haven't discussed yet. Because neutrinos only interact via the weak nuclear force, we are oblivious to them in everyday life. This is fortunate, because there are approximately sixty billion per square centimetre per second passing through your head from the nuclear reactions in the Sun. If the weak force were a little stronger, you'd get a hell of a headache. Actually, you wouldn't because you wouldn't exist, and this foreshadows the subject of the fine-tuning of the laws of nature we will undertake later in this chapter. The one remaining type of particle is the Higgs Boson, on its own in the fifth column. Empty space

THE POWER OF THE SUN
Without the weak nuclear force the Sun would not shine as it is an essential element in the Sun's energy production. We should be grateful it is still weak, otherwise life on Earth would be very uncomfortable.

ATLAS DETECTOR
ATLAS is one of six detector experiments being conducted at CERN, Geneva, at the Large Hadron Collider.

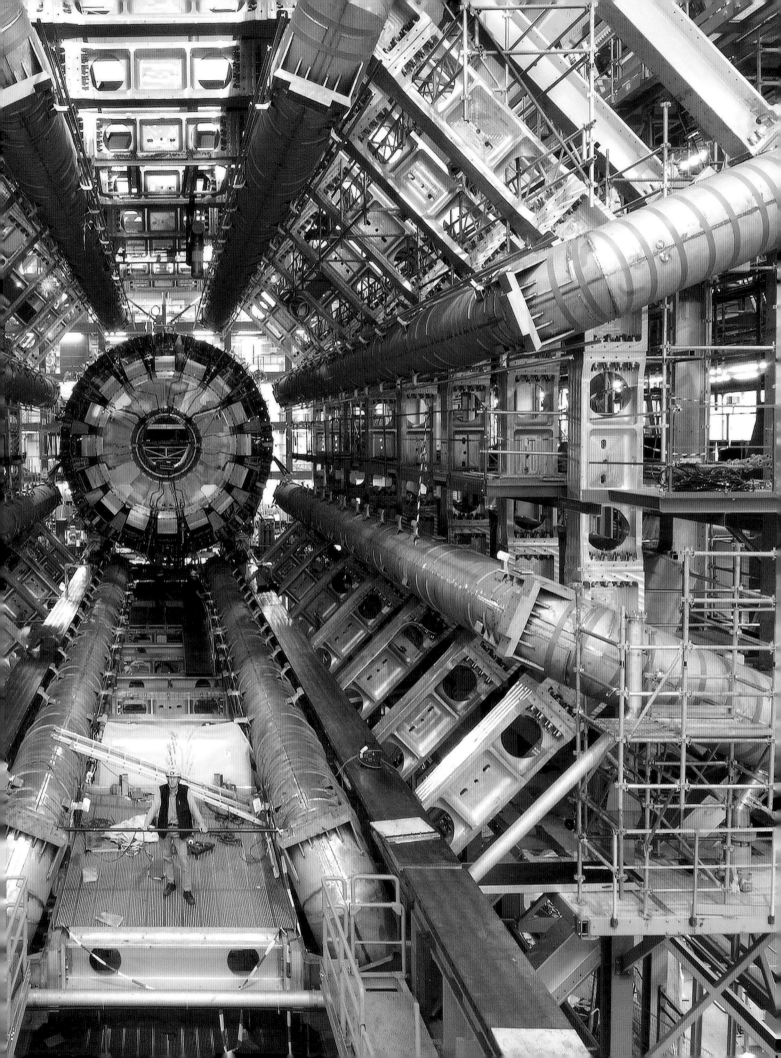

STANDARD MODEL

The Standard Model of particle physics is a theory that explains the interactions between sub-atomic particles in the form of the strong, weak and electromagnetic forces. The original theory has been tested experimentally since it was first postulated and has proven extremely robust. In 2013 the Higgs Boson that had been predicted by the theory was discovered using the Large Hadron Collider at CERN.

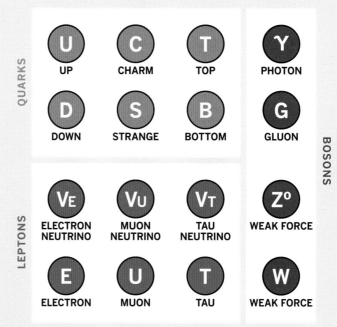

QUARKS

U	C	T
UP	CHARM	TOP
D	S	B
DOWN	STRANGE	BOTTOM

LEPTONS

V_E	V_U	V_T
ELECTRON NEUTRINO	MUON NEUTRINO	TAU NEUTRINO
E	U	T
ELECTRON	MUON	TAU

BOSONS

Y	PHOTON
G	GLUON
Z⁰	WEAK FORCE
W	WEAK FORCE

| H | HIGGS BOSON |

INSIDE THE ATOM

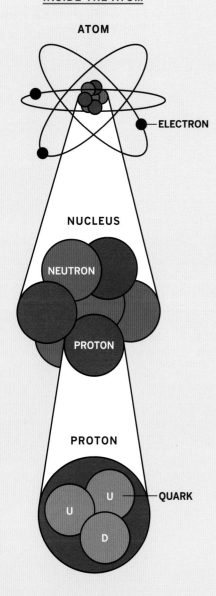

ATOM

ELECTRON

NUCLEUS

NEUTRON

PROTON

PROTON

QUARK

FUNDAMENTAL FORCES

		STRENGTH	RANGE (M)	PARTICLE
STRONG		1	10^{-15} (diameter of a medium-sized nucleus)	GLUONS
ELECTRO-MAGNETIC		$\frac{1}{137}$	INFINITE	PHOTON MASS = 0 SPIN = 0
WEAK		10^{-6}	10^{-8} (0.1% of the diameter of a proton)	INTERMEDIATE VECTOR BOSONS W^+ W^- Z^0 mass > 80 GeV spin = 1
GRAVITY		6×10^{-39}	INFINITE	Graviton ? mass = 0 spin = 2

isn't empty, but is jammed full of Higgs particles. All the known particles apart from the photon and the gluons, which are massless, interact with the Higgs particles, zigzagging through space and acquiring mass in the process. This is the counter-intuitive picture that was confirmed by the discovery of the Higgs Boson at CERN's Large Hadron Collider in 2012.

Two further generations of matter particles have been discovered. They are identical to the first generation except that the particles are more massive because they interact with Higgs particles more strongly. The muon, for example, is a more massive version of the familiar electron. The reason for their existence is unknown.

This is all there is in terms of the description of the fundamental ingredients of the universe. There are almost certainly other particles out there somewhere – the dark matter that dominates over normal matter in the universe by a factor of 5 to 1 is probably in the form of a new type of particle which we may discover at the Large Hadron Collider or a future particle accelerator. The evidence for dark matter is very strong and comes from astronomical observations of galaxy rotation speeds, galaxy formation models and the cosmic microwave background radiation that we met in Chapter 1 and will meet again later in this chapter. But because we don't know what form the dark matter takes, we are not able to incorporate it into our list.

The mathematical framework used to describe all the known particles and forces other than gravity is known as quantum field theory. It is a series of rules that allows the probability of any particular process occurring to be calculated. The whole thing can be described in one single equation, known as the Standard Model Lagrangian. Here it is:

$$L = -\frac{1}{4} W_{\mu\nu} W^{\mu\nu} - \frac{1}{4} B_{\mu\nu} B^{\mu\nu} - \frac{1}{4} G_{\mu\nu} G^{\mu\nu}$$

$$+ \overline{\psi}_j \gamma^\mu (i\delta_\mu - g\tau_j \cdot W_\mu - g' Y_j B_\mu - g_s T_j \cdot G_\mu) \psi_j$$

$$+ |D_\mu \phi|^2 + \mu^2 |\phi|^2 - \lambda |\phi|^4$$

$$- (y_j \overline{\psi}_{jL} \phi \psi_{jR} + y'_j \overline{\psi}_{jL} \phi_c \psi_{jR} + \text{conjugate})$$

It takes a lot of work to use this piece of mathematics to make predictions, but the predictions are spectacularly accurate and agree with every experimental measurement ever made in laboratories on Earth. This equation even predicted the existence of the Higgs particle; that's how good it is. It probably looks like a set of squiggles unless you are a professional physicist, but in fact it isn't too difficult to interpret, so let's dig just a little deeper. The 12 matter particles are all hidden away in the symbol ψ_j. The Standard Model is a quantum field theory because particles are represented by objects known as quantum fields. There is an electron field, an up quark field, a Higgs field and so on. The particles themselves can be thought of as localised vibrations in these fields, which span the whole of space. Fields will be important for us later, when we'll want to think about a certain type of field that may have appeared in the very early universe, known as a scalar field. The Higgs field is an example of a scalar field. The mathematical terms between the two ψ_js on the second line describe the forces and how they cause the particles to interact. The forces are also represented by quantum fields. The term $-g_s T_j \cdot G_\mu$ for example, describes the gluon field that allows the quarks in the ψ_j terms to bind together into protons and neutrons. The term g_s is known as the strong coupling constant. It is a fundamental property of our universe that encodes the strength of the strong nuclear force. Each of the forces has one of these coupling constants. We will want to discuss these coupling constants later, because they define what our universe is

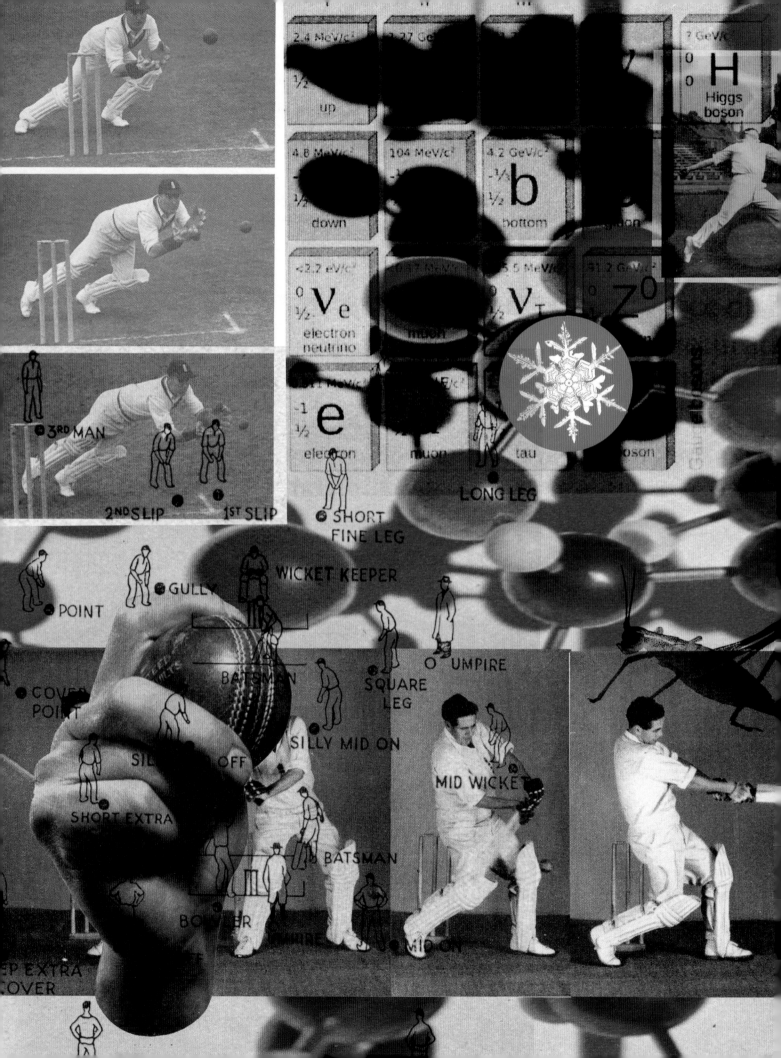

A GENIUS AT WORK
Einstein's manuscript for his
General Theory of Relativity
is a historic piece of
work-in-progress.

DARK MATTER
This makes up 26.8 per cent
of the total energy of the
observable universe and is not
described within the Standard
Model. It is likely that dark
matter will take the form of
a new type of particle, or
family of particles. There are
extensions to the Standard
Model, the simplest of which
is known as the Minimal
Supersymmetric Standard
Model, which are capable of
describing dark matter,
and we have no reason to
suspect that an entirely new
framework beyond quantum
field theory will be needed if
and when the nature of dark
matter is discovered.

like and what is allowed to exist within it. The last two lines deal with the Higgs Boson. The strength of the interaction between a matter particle and the Higgs field is contained in the y_j terms, which are known as Yukawa couplings. These must be inserted to produce the observed masses of the particles of matter. That's pretty much it.

Here ends our crash course on particle physics. The central point is that there exists a remarkably economical description of everything other than gravity, and it is contained within the Standard Model.

We considered the gravitational force in some detail in Chapter 1. It is described by Einstein's Theory of General Relativity, which is what physicists call a classical theory. There are no force-carrying particles in Einstein's theory; instead the force is described in terms of the curvature of spacetime by matter and energy and the response of particles to that curvature. A quantum theory of gravity, which we have already noted will be necessary to describe the first fleeting moments in the history of the universe, would involve the exchange of particles known as gravitons, but as yet nobody has worked out how to construct such a description. This is why Einstein's theory remains the only fundamental non-quantum theory we have.

For completeness, let's refresh our memory of Einstein's Theory of General Relativity:

$$G_{\mu\nu} = 8\pi G T_{\mu\nu}$$

General Relativity, like the Standard Model, contains a coupling constant encoding the measured strength of gravity: G, Newton's gravitational constant. The amount of dark energy is inserted by hand, in accord with observations, as was the case for the strengths of the forces and the masses of the particles in the Standard Model.

General Relativity and the Standard Model are the rules of the game. They contain all our knowledge of the way that nature behaves at the most fundamental level. They also contain almost all the properties of our universe that we think of as fundamental. The speed of light, the strengths of the forces, the masses of the particles (encoded as the strength of their interaction with the Higgs Bosons via the Yukawa couplings) and the amount of dark energy are all in these equations. In principle, any known physical process can be described by them. This is the current state of the art, but it doesn't mean that we know how everything works and can all retire, by a long shot or well-timed cover drive.

Most games are skin-deep, but cricket goes to the bone.
John Arlott and Fred Trueman

I timed a cover drive properly once when I was 14 years old playing at Hollinwood Cricket Club near Oldham. Front foot, head in line with the ball, sweet sound of the middle, four runs. I know what I have to do, but I never did it quite as well again. Cricket is an art built on simple rules, first codified by the members of the Marylebone Cricket Club on 30 May 1788; a significant date in world history according to historians with good taste. Those original laws still form the basis of the game today. There are 42 of them, and they define the framework within which each game evolves. Yet despite the rigid framework, no two games are ever alike. The temperature and humidity of the air, a light scatter of dew on the grass, the height of grass on the wicket, and hundreds of other factors will subtly shift and change throughout the game. More importantly, the players and umpires are each complex biological systems whose behaviour is far from predictable, with the exception of Geoffrey Boycott. The presence of so many variables makes the number of possible permutations effectively

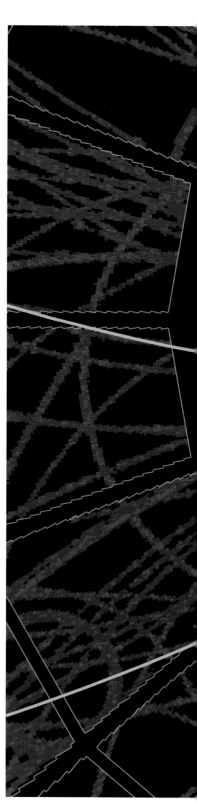

**PRODUCING
QUARK-GLUON PLASMA**
This computer simulation
reveals the type of particle
collision that occurs within
the Large Hadron Collider
which is expected to produce
quark-gluon plasma.

infinite, which is why cricket is the most interesting of human pursuits excluding science, sex and wine tasting.

Knowledge of the laws is therefore insufficient to characterise the infinite magic of the game. This is also true for the universe. The laws of nature define the framework within which things happen, but do not ensure that everything that can happen will happen in a finite universe – that rather obscure 'finite' caveat will be important for us later on. Virtually all of science beyond particle physics and theoretical

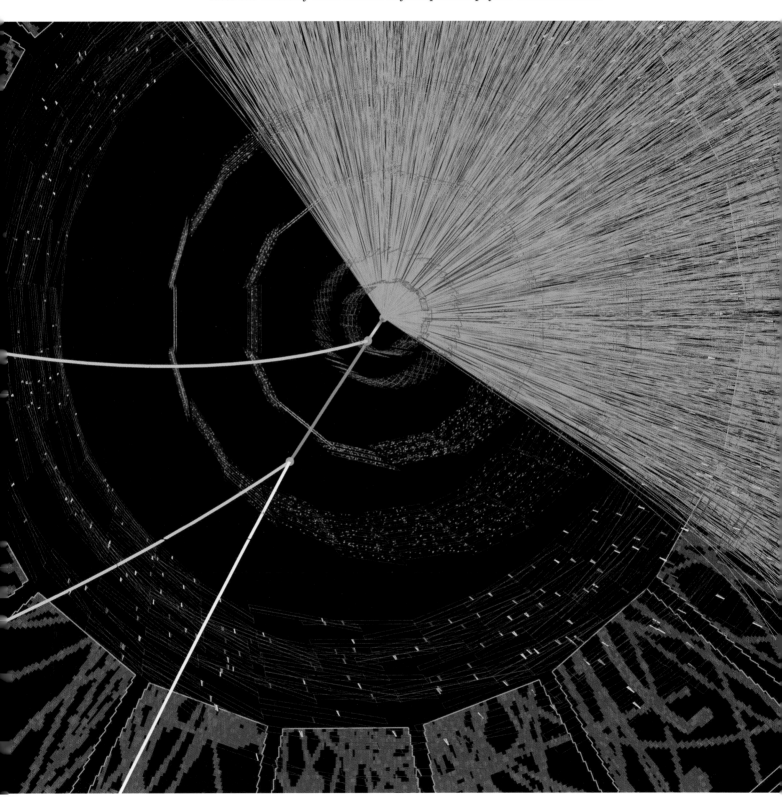

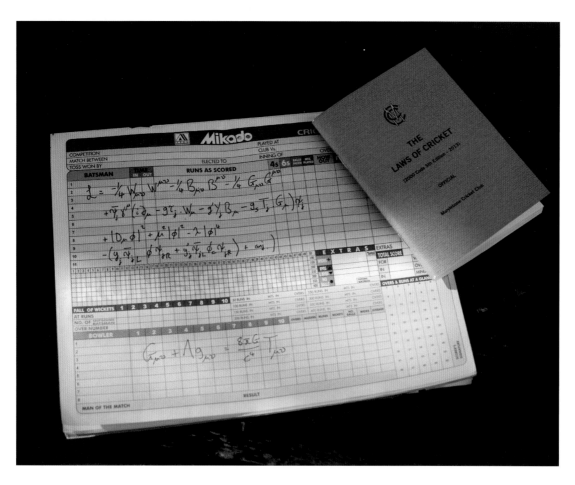

THE NATURE OF CRICKET
Just as in cricket, the laws of
nature have been set down
by watching the game of the
universe unfold, which is
what makes it, and the game
of cricket, so unique, and
ultimately so unpredictable.

cosmology is concerned with the complex outcomes allowed by the laws rather than the laws themselves, and in a certain sense our solipsistic initial question 'Why are we here?' is also a question about outcomes rather than laws. The answer to the question 'Why did England beat Australia in the great Ashes series of 2005?' is not to be found in the MCC rule book, and similarly the natural world that emerges from the Standard Model and General Relativity cannot be understood simply by discovering the laws themselves.

It's worth noting that the laws of nature were not written by the MCC, or even the committee of Yorkshire County Cricket Club. We had to work them out by watching the game of the universe unfold, which makes their discovery even more wonderful. Imagine how many matches would have to be viewed in order to deduce the laws of cricket, including but not restricted to the Duckworth Lewis method? The great achievement of twenty-first-century science is that we've managed to work out the laws of nature by doing just this; observing many millions of complex outcomes and working out what the underlying laws are.

The Standard Model, then, cannot be used to describe complex emergent systems such as living things. No biologist would attempt to understand the way that ATP is produced inside cells using the Standard Model Lagrangian and no telecommunications engineer would use it to design an optical fibre. They wouldn't want to even if they could; you wouldn't gain any insight into how a car engine works by starting off with a description of its constituent subatomic particles and their interactions. So whilst it is important that we have a detailed model of nature at the level of the known fundamental building blocks, we must also understand how the complexity we observe around us emerges from these simple laws if we are to make progress with our difficult 'Why?' question.

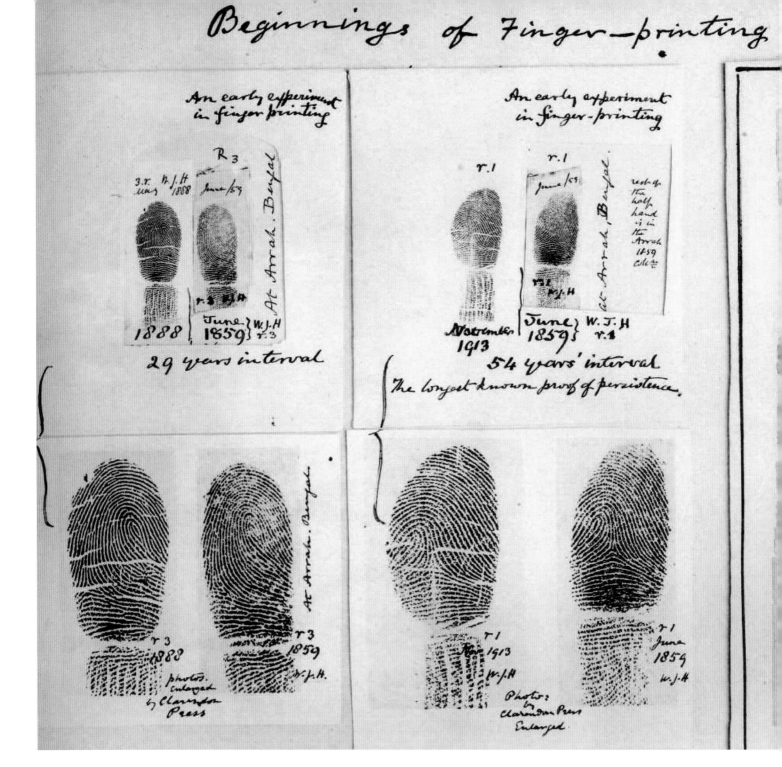

Beginnings of Finger-printing

An early experiment in finger-printing

An early experiment in finger-printing

29 years interval

54 years' interval

The longest known proof of persistence.

NATURE'S FINGERPRINT

When you have eliminated the impossible, whatever remains, however improbable, must be the truth.
Sherlock Holmes

On Monday 27 March 1905 at 8.30am, William Jones arrived at Chapman's Oil and Colour Shop on Deptford High Street ready for a day's work. Jones normally arrived a few minutes after the shop manager Thomas Farrow had raised the shutters. On this particular Monday, however, the shutters were down. Farrow lived with his wife Anne above the shop, but no matter how hard Jones knocked on their door,

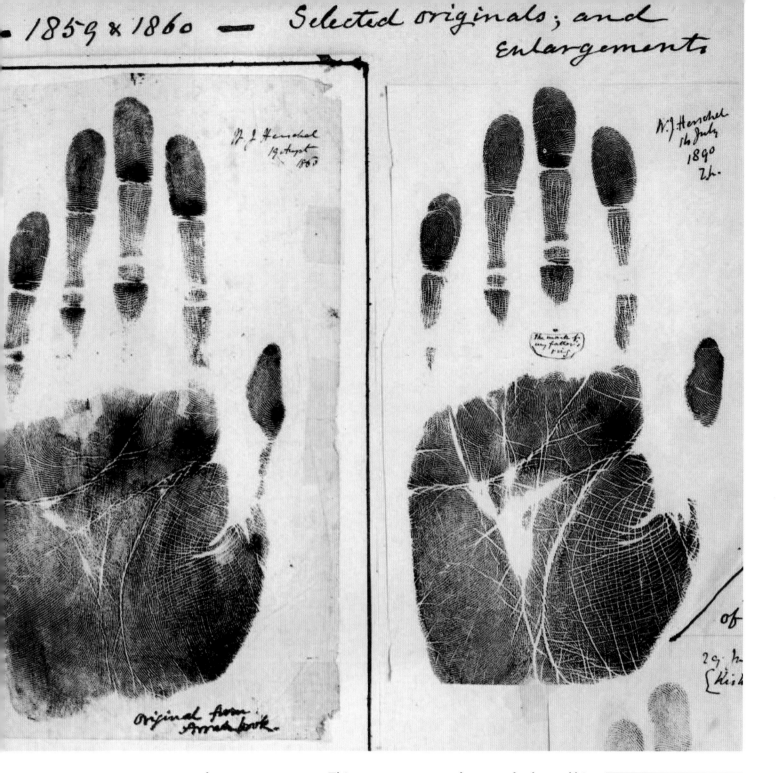

there was no response. This was a most unusual start to the day, and his concern increased when a glimpse through a window revealed chairs strewn across the floor of the normally tidy shop. Jones and another local resident forced the door, to be confronted by Farrow lying dead in a pool of blood. Anne had been similarly bludgeoned in her bed, although she clung to life for four more days without regaining consciousness.

Such scenes were not uncommon in Edwardian London. The reason that this crime is of note is because it was the first in the world to use a new technology to catch and convict the killers. On an inner surface of the empty cash box, the police noticed a fingerprint. They already had a suspect; a local man named Alfred Stratton, who was arrested three days later along with his brother Albert. The Strattons' fingerprints were taken, and a positive match was made between the cash-box print and

COMPLEX CLUES OF THE UNIVERSE
The uncovering of the uniqueness of fingerprints changed the face of police work in the early twentieth century. Although fingerprints appear intricate and complex, the process by which they were formed is far simpler.

Alfred Stratton's right thumb. Although never used before in a murder case, expert witnesses convinced the jury that the complex patterns of the cash-box fingerprint could only belong to Alfred Stratton. The jury took just two hours to find the Stratton brothers guilty of murder, and the pair were sentenced to death by hanging, with justice swiftly dispatched on 23 May.

Take a look at your fingerprints now; there is seemingly endless complexity in the swirls and ridges. Since every human being carries different fingerprints on the hands and the soles of their feet (which aren't fingerprints, but there isn't a word for them), the size of database required to characterise every human being's fingerprints would be colossal. One of the most important properties of nature, however, is that the blueprints for the construction of the natural world are far simpler than the natural world itself. In modern language, there is a tremendous amount of data compression going on. The instructions to create fingerprints are far simpler than the fingerprints themselves, and more than that, the same instructions, run over and over again from slightly different starting points in the embryonic stage of our development, always lead to different fingerprints. This behaviour shouldn't come as a

SHIFTING SANDS
The formation of sand dunes and patterns within deserts may appear arbitrary but is in fact the product of simple laws which govern their movement.

surprise. The sweep of desert dunes or the patterns in summer clouds are all described by a handful of simple laws governing how sand grains or water droplets behave when agitated by shifting air currents, buffeted by chaotic thermals and winds and re-ordered by the action of the forces of nature. And yet from a simple recipe, complexity emerges.

The quest to understand how the boundless variety of the natural world emerges from underlying simplicity has been a central theme in philosophical and scientific thought. Plato attempted to cast the world available to our senses as the distorted and imperfect shadow of an underlying reality of perfect forms, accessible through reason alone. The modern expression of Plato's ethereal dualism was captured eloquently by Galileo, 500 years ago: 'The book of nature is written in the language of mathematics'. The challenge is not only to discern the underlying mathematical behaviour of the world, but also to work back upwards along the chain of complexity to explain how those forms that Plato would have defined as imperfect arise from the assumed lower-level perfection. A rather beautiful early example of this quest is provided by Galileo's illustrious contemporary, Johannes Kepler.

A STUDY IN SKIN
Each fingerprint reminds us that although they originate from simple starting points, their development will always lead to different results.

A BRIEF HISTORY OF THE SNOWFLAKE

As I write it has begun to snow, and more thickly than a moment ago. I have been busily examining the little flakes.
Johannes Kepler

Johannes Kepler is rightly best known for his laws of Planetary Motion that paved the way for Newton to write *Principia*. Hidden within his illustrious CV, however, is a publication that had a rather more whimsical earthbound ambition. Two years after publishing the first part of *Astronomia Nova* in 1609, Kepler published a short 24-page paper entitled *De nive sexangula – On the Six-Cornered Snowflake*. It is a beautiful example of a curious scientific mind at work. In the dark December of 1610, Kepler was walking across the Charles Bridge in Prague when a snowflake fell on the lapel of his coat. In the freezing night he stopped and wondered why this ephemeral sliver of ice possessed a six-sided structure, in common with all other snowflakes, notwithstanding their seemingly infinite variation (see next page). Others had noticed this symmetry before, but Kepler realised that the symmetry of a snowflake must be a reflection of the deeper natural processes that underlie its formation.

'Since it always happens when it begins to snow, that the first particles of snow adopt the shape of small six-cornered stars, there must be a particular cause,' wrote Kepler, 'for if it happened by chance, why would they always fall with six corners and not with five, or seven?' Kepler hypothesised that this symmetry must be due to the nature of the fundamental building blocks of snowflakes. This stacking of frozen 'globules', as he referred to it, must be the most efficient way of building a snowflake from the 'smallest natural unit of a liquid like water'.

To my mind, this is a leap of genius and a tremendously modern way of thinking about physics. The study of symmetry in nature lies at the very heart of the Standard Model, and abstract symmetries known as gauge symmetries are now known to be the origin of the forces of nature. This is why the force-carrying particles in the Standard Model are known as gauge bosons. Kepler was searching for the atomic structure of snow before we knew atoms existed, motivated by the observation of a symmetry in nature – the six-sided shape of all snowflakes. The inspiration for this idea, which is way ahead of its time, came from a peculiar source. In the years leading up to the publication of *De nive sexangula*, Kepler had been in communication with Thomas Harriot, an English mathematician and explorer. Amongst multiple claims to fame, Harriot was the navigator on one of Sir Walter Raleigh's voyages to the New World, and had been asked to solve a seemingly simple mathematical problem. Raleigh wanted to know how best to stack cannonballs to make the most efficient use of the limited space on the ship's deck. Harriot was driven to exploring the mathematical principles of sphere packing, which in turn led him to develop an embryonic model of atomic theory and inspire Kepler's consideration of the structure of snowflakes. Kepler imagined replacing cannonballs with globules of ice, and supposed that the most efficient arrangement creating the greatest density of globules was the six-sided hexagonal form he observed in the snowflake on his shoulder. Kepler also observed hexagonal structures across the natural world, from beehives to pomegranates and snowflakes, and presumed that there must be some deeper reason for its ubiquity.

'Hexagonal packing', as Kepler referred to it, must be 'the tightest possible, so that in no other arrangement could more pellets be stuffed into the same container'. This became known as the Kepler Conjecture. It took almost 400 years to prove Kepler's conjecture, and this required the help of a 1990s supercomputer. Despite the time lag, Kepler's work

THE SYMMETRY OF SNOWFLAKES
A snowy walk in Prague led Johannes Kepler to formulate what became known as the Kepler Conjecture, based on the symmetrical, hexagonal structure of the snowflakes that settled around him.

had a more immediate impact, inspiring the beginnings of modern crystallography that led eventually to the discovery of the structure of DNA. What a lovely example of serendipity coupled with curiosity and a sprinkling of genius; from cannonballs to snowflakes to the code of life.

As for Kepler's original thought on that frozen bridge, he never found the connection between the underlying structure of his ice globules and the hexagonal symmetry of snowflakes. Even though he realised that the regular patterns must reveal something about the shape of the building blocks of snowflakes and the details of the packing, he couldn't explain the ornate complexity or the flatness of the structure. Instead he acknowledged his failure with the good grace of a true scientist: 'I have knocked on the doors of chemistry' he writes at the end of his paper, 'and

Philos. Trans. Vol. XLIX. TAB. XXI. *p.* 647.

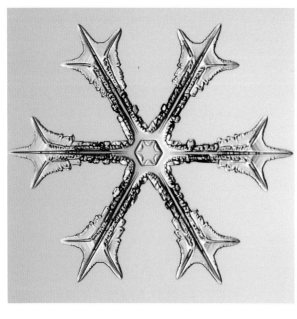

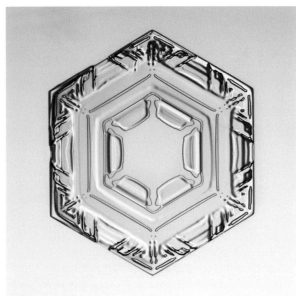

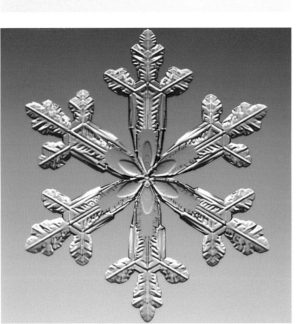

seeing how much remains to be said on this subject before we know the cause, I would rather hear what you think, my most ingenious man, than wear myself out with further discussion.'

Three and a half centuries later, Japanese physicist Ukichiro Nakayara made the first artificial snowflakes in a laboratory. Writing in 1954, he describes a process that begins not with the snowflake itself but with smaller substructures called snow crystals, which are in turn built up from collections of ice crystals – the globules Kepler was searching for. The hexagonal packing that Kepler suspected to be the origin of the snowflakes' symmetry begins with the formation of these ice crystals, when water molecules link together in a hexagonal structure via hydrogen bonds. Hydrogen bonding occurs because of the structure of the water molecules themselves, with a greedy oxygen atom hungry for electrons grabbing them off two hydrogen atoms, forming covalent bonds that lock the H_2O molecules together, leaving a residual positive electrical charge in the vicinity of the two protons and a negative charge in the vicinity of the oxygen. This slight separation of charge in the water molecules allows them to bind together into larger structures through the mutual attraction and repulsion of the electrical charges, just as an electron is bound into its position around an atomic nucleus. The entire configuration, including the structure of the oxygen nucleus and the single protons that comprise the hydrogen nuclei, can be predicted in principle by the Standard Model of particle physics. Yet the details of any particular snowflake are beyond computation, because the seemingly infinite variety reflects the precise history of the snowflake itself. Once ice crystals form as agglomerations of water molecules held together by hydrogen bonds, they cluster around dust particles in the air, building on their underlying hexagonal symmetry to form larger snow crystals. As the crystals begin the long journey down to Earth they join in ever-larger, more complex combinations, shaped by endless variations of air temperature, wind patterns and humidity into myriad unique forms. The symmetry is all that remains of the simplicity, and it takes a careful and patient eye to see the endless variation for what it is; a reflection of the complex history of the snowflake convoluted with the underlying simplicity of the laws of nature.

The most vivid example of emergent complexity, and the closest to our hearts, is life. As we discussed in Chapter 2, the origin of life on Earth has a sense of inevitability about it, because its basic processes are chemical reactions that will proceed given the right conditions. Those conditions were present in the oceans of Earth 3.8 billion years ago, possibly earlier, and they led to the emergence of single-celled organisms. The fateful encounter which produced the eukaryotic cell around 2 billion years ago looks rather more like blind chance, but it happened here and laid the foundations for the Cambrian explosion 530 million years ago. There is a bit of hand-waving going on here, though, and to make a more persuasive case that all the complexity of Darwin's endless forms most beautiful can at least in principle emerge from simple underlying laws, one more example is in order.

Perhaps the most beautiful manifestation of the artful complexity of nature can be found in the spots, stripes and patterns on the coats and skin of living things; emergent pattern writ large across venomous striped surgeonfish, emperor angelfish, zebra swallowtail butterflies and the big cats of Africa and Asia. Everyone agrees that these patterns evolved as a result of natural selection of one form or another, and the raw material for the variation was provided by random mutations in the genetic code. But a very challenging scientific question of fundamental importance in modern biology is precisely how patterns such as these appear.

NEW PERSPECTIVES
Johannes Kepler got the scientific world talking about and observing snowflakes in a whole new light. English scientist Robert Hooke published his sketches and observations in *Micrographia: or some physiological descriptions of minute bodies made by magnifying glasses, with observations and enquiries thereupon* in 1665.

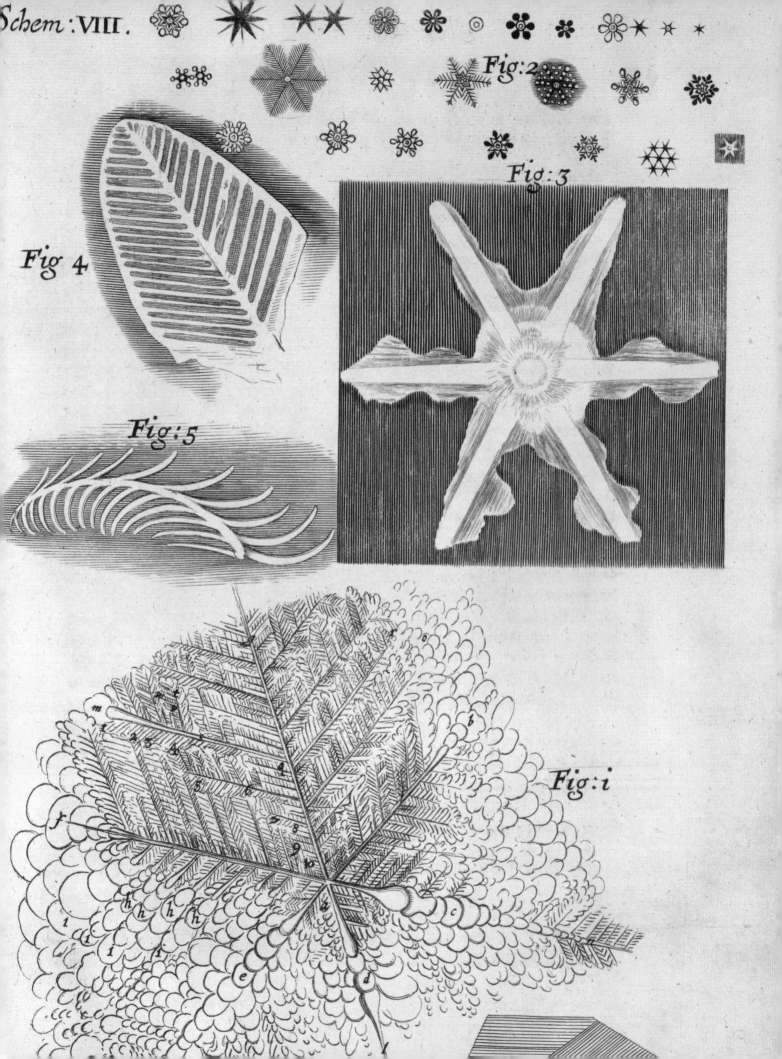

Schem: VIII.

Fig: 2

Fig: 3

Fig: 4

Fig: 5

Fig: 1

ART IN NATURE

The natural world illustrates its complexity even through the simplest observation of appearance. Patterns reveal themselves through spots and stripes on animal hides and skins and in the composition of plants and their fruits, as in this Romanesco broccoli.

HOW THE LEOPARD GOT ITS SPOTS

...Zebra moved away to some little thorn-bushes where the sunlight fell all stripy and the Giraffe moved off to some tallish trees where the shadows fell all blotchy.
'Now watch,' said the zebra and the giraffe. 'This is the way it's done. One... two...three! And where's your breakfast!' ... All they could see were stripy shadows and blotched shadows in the forest, but never a sign of Zebra and Giraffe.
'That's a trick worth learning. Take a lesson from it, Leopard!'
...Then the Ethiopian put his five fingers close together and pressed them all over the leopard, and wherever the five fingers touched, they left five black marks, all close together...
Rudyard Kipling

Rudyard Kipling's *Just So* story, 'How The Leopard Got His Spots', tells the story of an Ethiopian man and a leopard. They went hunting together, but one day the man noticed that the leopard wasn't very successful. The reason, he deduced, was that the leopard had a plain sandy coat, whereas all the other animals had camouflage. 'That's a trick worth learning, leopard' he said, taking his fingers and thumb and pressing them into the leopard's coat to give it the distinctive five-pointed pattern. If you don't believe in evolution by natural selection, this is the most plausible theory open to you. If you do, then what remains is to identify the mechanism by which the pattern is formed. The answer might appear to be solely a matter of genetics, but genes are not the whole story. It would take a terrific amount of information to instruct every single cell to colour itself according to its position on the leopard's skin, and this is indeed not what is done. Nature is frugal and deploys a much more efficient mechanism for producing camouflage patterns. As with so many things in this book, I get to say yet again that this is an active area of research, and therefore exciting. The reason for the attention is that camouflage patterns on the skin self-organise during the development of the embryo, and embryonic development is of course fundamental to an understanding of biology. In the case of the leopard, it is thought, though not proven, that the camouflage is an example of a Turing pattern, named after the great Bletchley Park code-breaker and mathematician Alan Turing.

In 1952, Turing became interested in morphogenesis – the process by which an animal develops its shape and patterning. He was particularly interested in the mathematics behind regularly repeating patterns in nature such as the Fibonacci numbers and golden ratio in the leaf arrangements of plants and the scales of pineapples, and the appearance of camouflage patterns such as the tiger's stripes and the leopard's spots. Turing's influential and ground-breaking paper, 'The Chemical Basis of Morphogenesis', published in March 1952, begins with a simple statement. 'It is suggested that a system of chemical substances, called morphogens, reacting together and diffusing through a tissue, is adequate to account for the main phenomena of morphogenesis.' These systems are known as reaction-diffusion systems, and they can produce patterns from a featureless initial mixture if the two reactants diffuse at different speeds. There is a nice analogy that describes how such a system can work. Imagine a dry field full of grasshoppers. They are strange grasshoppers, because when they get warm they sweat, generating a large amount of moisture. Now imagine that the field is set alight in several different places. The flames will spread at some fixed speed, and if there were no grasshoppers the whole field would be charred. As the flames approach the grasshoppers, however, they will start to sweat, dampening the grass behind them and inhibiting the flames as they hop away ahead of the approaching flames. Depending on the different parameters, including the different speeds of the flames and the grasshoppers, and the amount of sweat necessary to quell the advancing flames, a Turing pattern can be formed, with areas of charred grass and green areas where the inhibiting grasshoppers prevented the fire from taking hold.

It is thought that the leopard gets its spots in this way during embryonic development: an activator chemical (fire) spreads through the skin and stimulates the production of the dark pigmented spots (charred grass) but is inhibited by another chemical (sweating grasshoppers) spreading with a higher diffusion rate. The precise pattern produced depends on the 'constants of nature' of the system, such as the speeds at which the chemicals diffuse, and on what a mathematician would call the boundary conditions: the size and geometry of the grassy field in our analogy. In embryonic development, it is the size and shape of the embryo when the reaction-diffusion begins that determines the type of

PICK A PATTERN
The distinctive pattern on the leopard's coat – like that of so many other wild animals – is thought (although still unproven) to be an example of a Turing pattern, the theory of how animals develop their patterning that was put forward by Bletchley Park code-breaker Alan Turing in 1952.

GRASSHOPPERS
Published by J. D. Murray in *Notices of the AMS*, June/July 2012: 'Why are there no 3-headed monsters? Mathematical modelling in biology'.

pattern produced. A long and thin domain produces stripes. A domain that is too small or too large produces uniform colour. In between can be found the distinctive coat patterns of cows, giraffe, cheetah and, of course, the leopard. Computer simulations of Turing patterns have been remarkably successful, not only in describing the generic features, particularly of mammalian coats, but also some of the interesting details seen in nature. For example, the mathematical models predict that it is possible for spotted animals to have stripped tails, as cheetahs do, but not for striped animals to have spotted tails; and indeed, no such examples exist.

Kepler's snowflakes and the leopard's spots are two picturesque examples of emergent complexity: the appearance of intricate, ordered patterns from the action of simple underlying laws. Nature contains systems far more complex than these, of course: you being a case in point. But to return to the question at the beginning of our solipsistic meander, the reason that you exist, given the laws of nature, is that you

CHEMICAL WAVES
The chemical waves in this solution demonstrate the theory that just a tiny alteration in conditions or contents can cause an altered image, such as a dust particle on the surface or the level of moisture in the air.

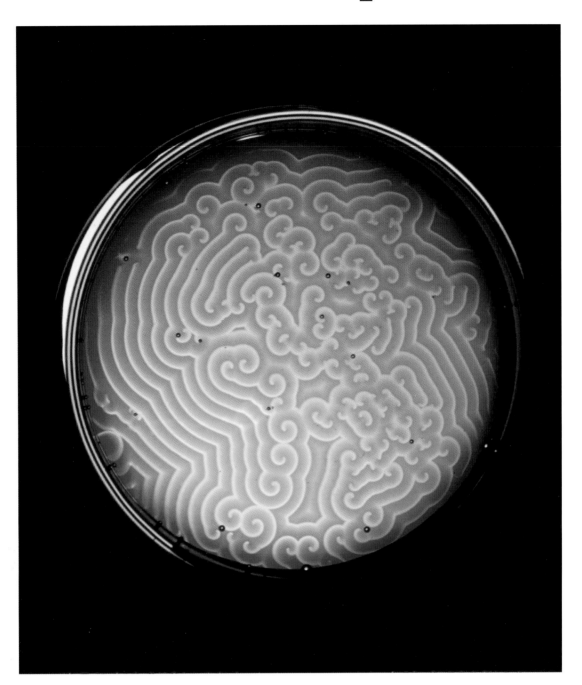

are allowed to. Just as all snowflakes and all leopards' coats are unique in detail because of their individual formation histories, so you are unique because no two human beings share a common history. But we wouldn't read any deep meaning into the existence of one particular snowflake in a snowstorm, and the same is true for you. Our focus should therefore shift from trying to explain the appearance of humans, or our planet, or even our galaxy, to a rather deeper question: the origin of the whole framework – of spacetime and the laws that govern it and the allowed structures within it. What properties of the laws themselves are essential for galaxies, planets and human beings to exist? After all, as we've noted, the laws might be mathematically elegant and economical, but they do contain a whole host of seemingly randomly chosen numbers, discovered by experimental observation and with no known rhyme or reason to them – the constants of nature such as the strengths of the forces, the masses of the particles and the amount of dark energy in the universe. How dependent is our existence on these fundamental numbers?

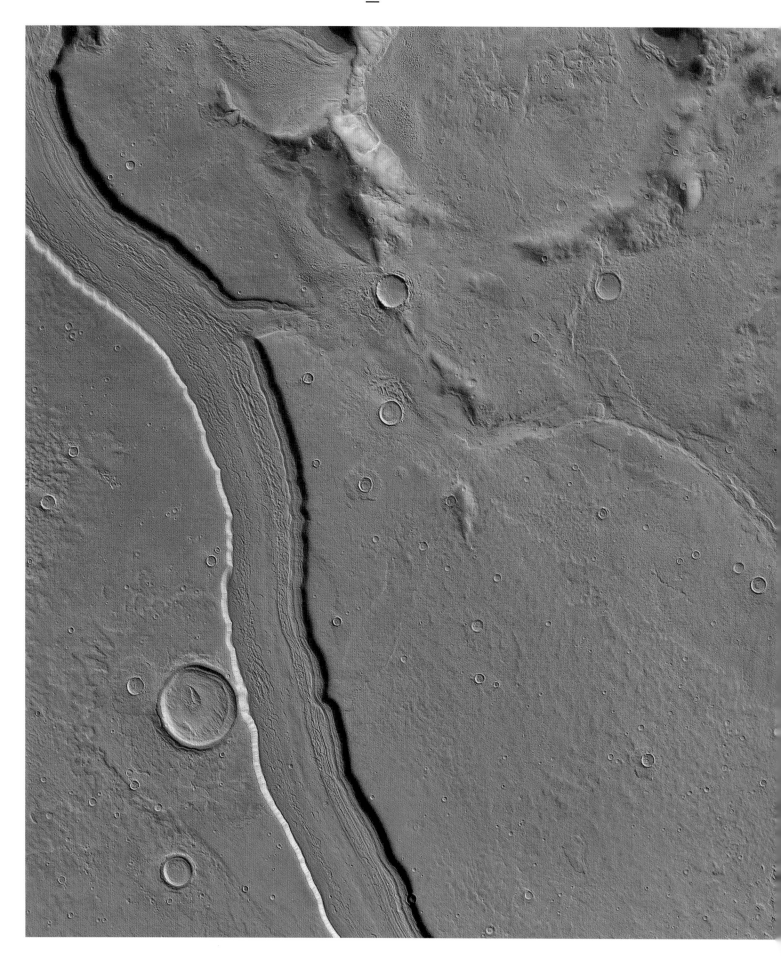

WHY ARE WE HERE?

**OUTLINES OF
ANOTHER WORLD**
Ongoing research hopes
to reveal that nature on all
planets within the galaxies
plays a major part in shaping
their landscapes. This image,
taken by the ESA Mars Express
Orbiter on 14 May 2012,
reveals a surface texture and
a river-like channel that have
been sculpted by flowing
water sometime in the planet's
ancient past.

A UNIVERSE MADE FOR US?

Our universe appears to be made for us. We live on a perfect planet, orbiting around a perfect star. This is of course content-free whimsy. The argument is backwards. We have to be a perfect fit for the planet because we evolved on it. But there are interesting questions when we look deeper into the laws of nature and ask what properties they must have to support a life in the universe. Take the existence of stars, for example. Stars like the Sun burn hydrogen into helium in their cores. This process involves all four forces of nature working together. Gravity kicks it all off by causing clouds of dust and gas to collapse. As the clouds collapse, they get denser and hotter until the conditions are just right for nuclear fusion to occur. Fusion starts by turning protons into neutrons through the action of the weak nuclear force. The strong nuclear force binds the protons and neutrons together into a helium nucleus, which in itself exists on account of the delicate balance between the strong nuclear force holding it together and the electromagnetic force trying to blow it apart because of the electrically charged protons. When stars run out of hydrogen fuel, they perform another series of equally precarious nuclear reactions to build carbon, oxygen, and the other heavy elements essential for the existence of life. What happens if the strengths of the forces, those fundamental constants of nature we met earlier in the chapter, are varied a bit?

There are many examples of apparent fine-tuning in nature. If protons were 0.2 per cent more massive, then they would be unstable and decay into neutrons. That would certainly put an end to life in the universe because there would be no atoms. The proton mass is ultimately set by the details of the strong and electromagnetic forces, and the masses of the constituent quarks, which are set by the Yukawa couplings to the Higgs field in the Standard Model. There really isn't much freedom at all.

The mother of all fine-tunes, however, is the value of our old friend dark energy, the thing that is causing our universe to gently accelerate in its expansion. Although dark energy contributes 68 per cent of the energy density of the universe, the amount of dark energy in a given volume of space is actually small. Very small: 10^{-27}kg per cubic metre to be precise. The point is that every cubic metre of our universe has this amount of dark energy in it, and that adds up! Explaining why dark energy has this small, but non-zero, value is one of the great problems in cosmology, not least because if a particle physicist sits down with quantum field theory and decides to calculate how big it should be, it turns out that it would be more naturally of the order of 10^{97}kg per cubic metre. That's a lot bigger than 10^{-27}kg per cubic metre. Over a million times bigger, in fact. That's embarrassing for the particle physicists, of course, but from the perspective of fine-tuning it's even worse. If the value of dark energy were only 50 times larger than it is in our universe, rather than somewhere else in this immensely large theoretical range, then it would have become dominant in the universe around one billion years after the Big Bang during the time that the first galaxies were forming. Because dark energy acts to accelerate the universe's expansion and dilute matter and dark matter, gravity would have lost the battle in such a universe and no galaxies, or stars, or planets or life would exist. What could possibly account for this incredible piece of luck? It can't really be luck – the odds are too long by a Geoffrey Boycott innings. One possibility is that there is some as yet unknown physical law or symmetry that guarantees that the amount of dark energy will be very close to, but not quite, zero. This is certainly possible, and there are physicists who believe that this may be the case. The other possibility, which was raised by one of the fathers of

the Standard Model, Steven Weinberg, is that the value of dark energy is anthropically selected. Anthropic arguments appear at one level to be a statement of the obvious: the properties of the universe must be such that human beings can exist because human beings do exist. This is, of course, true, but it is fairly devoid of content from a physical perspective *unless* there is some way in which all possible values of dark energy, and indeed all the other constants of nature, are realised somewhere. If, for example, there exists a vast, possibly infinite swathe of different domains in the universe, or indeed an infinity of other universes, each with a different amount of dark energy selected by some mechanism from the span of allowed values, then we would indeed have a valid anthropic explanation

DARK MATTER DETECTOR
The LUX Dark Matter Detector.

for our 'special' human universe. It must exist, because they all do, and of course we appear in the one that permits our existence.

But surely it makes no sense to take refuge in a vast infinity of universes to explain our existence? Absolutely correct, if that's why the idea is introduced – it's no better than a God-of-the-gaps explanation. If, however, there were some other reason, based on observations and theoretical understanding, that suggested an infinity of universes, then such an anthropic explanation for our perfect, human universe would be admissible. Remarkably – and that remarkably overused word is appropriate for once – this outlandish suggestion is a widely held view amongst many cosmologists.

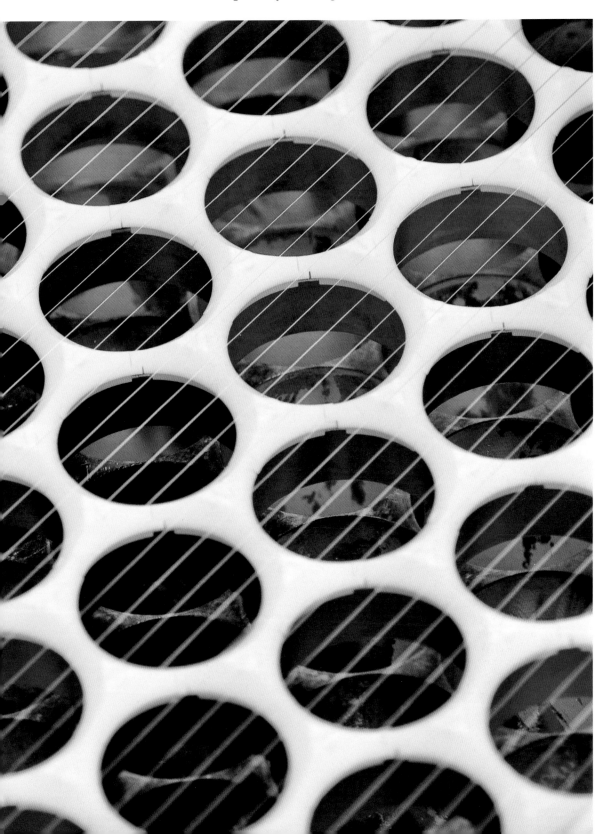

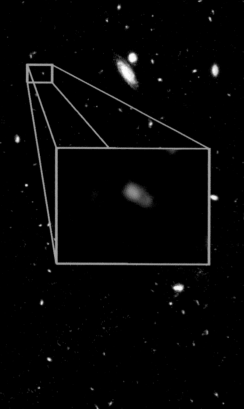

A DAY WITHOUT YESTERDAY?

*It suddenly struck me that that tiny pea, pretty and blue, was the Earth.
I put up my thumb and shut one eye and my thumb blotted out the planet
Earth. I didn't feel like a giant. I felt very, very small.*
Neil Armstrong

If we look at our universe on the largest distance scales, by which I mean at distance scales far larger than the size of single galaxies, it has a number of properties that any theory of its origin has to explain. The most precise picture of the young universe we have is the photograph of the Cosmic Microwave Background Radiation (CMB) taken by the Planck satellite.

This is the afterglow of the Big Bang, a photograph of the universe as it was 380,000 years after the initial hot, dense phase when the expansion had cooled things down sufficiently for atoms to form. The most obvious feature of the CMB is that it is extremely uniform, glowing at a temperature of 2.72548 degrees above absolute zero, with small fluctuations at the level of 1 part in 100,000. Those very tiny temperature differences are represented by the colours in the photograph. This uniformity is extremely difficult to explain in the standard Big Bang model for a simple reason. Our observable universe today is 90 billion light years across. This means that if we look out to the CMB from opposite sides of the Earth, we are looking at two glowing parts of the ancient sky that are now separated by 90 billion light years. The universe, however, is only 13.8 billion years old, which means that light, the fastest thing there is, has only had time to travel 13.8 billion light years. Two 'opposite' parts of the CMB could therefore never have been in contact with each other in the standard Big Bang model, and there is absolutely no reason why they should be *almost* precisely the same temperature. I've italicised 'almost' in the previous sentence because, as we noted, there are very slight variations in the CMB at the level of 1 part in 100,000, and these are very important. The universe was never completely smooth and uniform everywhere, and these variations in density are encoded into the CMB as differences in temperature. The regions of slightly greater density ultimately seeded the formation of the galaxies, and so without them we wouldn't exist. What caused these small variations in the otherwise ultra-smooth early universe?

Another fundamental property of the universe that is difficult to explain is its curvature – or lack of it – which can also be measured from the CMB. Space appears to be absolutely flat; a veritable ice rink. Recall from Chapter 1 that the shape of space is related to the density and distribution of matter and energy in the universe through Einstein's equations. In the standard Big Bang theory, the universe doesn't have to be flat. In fact, it requires a great deal of fine-tuning to keep it flat over 13.8 billion years of cosmic evolution. Instead, the radius of curvature is measured to be much greater than the radius of the observable universe – more than sixty orders of magnitude larger. That's a big problem!

In the early 1980s, the need to explain these and other properties of the observable universe led a group of Russian and American physicists to propose a radical idea. The modern version, the best-known proponents of which are Alan Guth, Andrei Linde and Alexei Starobinsky, is known as the Theory of Inflation. We'll describe a particular version of inflation below, driven by something called a scalar field, which was first described by Andrei Linde.

Spacetime existed before the Big Bang, and for at least some of that time was described by Einstein's Theory of General Relativity and a quantum field theory like the Standard Model. The central idea in quantum theory is that anything that can happen does happen.

Everything that is not explicitly ruled out by the laws of nature will happen, given enough time. One of the types of things permitted to exist in quantum field theory are scalar fields. We've met an example of a scalar field earlier in the chapter in the guise of the Higgs field, which we know to exist because we've measured it at the Large Hadron Collider. Scalar fields have the property that they can cause space to expand exponentially fast. We touched on such a scenario in Chapter 1 without being explicit about the mechanism – it is the de Sitter's matter-less solution to Einstein's field equations first discovered in 1917. Given general relativity and quantum field theory, therefore, it must be the case that scalar fields will fluctuate into existence in such a way that an exponential expansion of spacetime is triggered. In this exponential phase, spacetime expands faster than the speed of light. This might sound problematic if you know some relativity, but it isn't. The universal speed limit exists for particles moving through spacetime, but does not apply for the expansion of spacetime itself. In a tiny fraction of a second – around 10^{-35} seconds in fact – an exponential expansion of this type can inflate a piece of spacetime as tiny as the Planck length to a quite mind-boggling size: trillions of times larger than the observable universe. Any pre-existing curvature is completely washed out, leading to a flat observable universe. It's like looking at a square centimetre-sized piece of the surface of a balloon of a light year in radius; you won't see any curvature, no matter how hard you try.

Likewise any variations in density will be washed out, leading to the smooth and uniform appearance of the CMB. Perhaps the greatest triumph of inflationary models such as these, however, is that they don't predict a completely uniform, homogeneous and isotropic universe. Quantum theory doesn't allow for absolute uniformity. Empty space is never empty, but a fizzing, shifting soup of all possible quantum fields. Like the surface of a stormy ocean, waves in the fields are constantly rising and falling, and the exponential expansion can freeze these undulations into the universe. Remarkably, when calculations using the known laws of quantum theory are carried out, the sort of density fluctuations that result from such a mechanism are precisely of the form seen in the CMB. These quantum fluctuations are the seeds of the galaxies and therefore the seeds of our existence, frozen into the oldest light in the cosmos and photographed by a satellite built by the people of Earth 13.8 billion years later.

Inflation in this guise explains the observable properties of our universe, and in particular all the details of the CMB, which has been measured to high accuracy. This is why it is currently widely accepted as an essential ingredient by many cosmologists. As if this wasn't enough to get excited about, however, there is much more.

One obvious question that arises is this: if inflation gets going, how does it stop? The answer is that inflation stops completely naturally, but with a fascinating twist that drives right to the heart of our 'Why are we here?' question. The scalar field driving inflation fluctuates up and down in accord with the laws of quantum theory, just like the waves on the surface of an ocean. If the energy stored in the field is high enough, inflation begins. One might expect that such a rapid expansion would dilute the energy extremely rapidly, causing inflation to stop. But scalar fields have the interesting property that their energy density can stay relatively constant as space expands. You can think of the expanding space as doing work on the field, pumping energy into it and keeping its level high. And in turn, the high level of the field's energy continues to drive the expansion. This might sound like the ultimate free lunch, and in a sense it is, almost, although gradually the energy will become diluted and decay away. The time this takes depends on the size of the initial fluctuation in the field and the details of the field itself, but in general

INFLATION IN ACTION
The sport of zorbing brilliantly illustrates the theory of inflation; putting in action the analogy of a very large ball rolling down a slope!

the higher the initial energy, the longer the field takes to fall in value as the expansion continues. An analogy often used to picture this scenario is to imagine a ball rolling down the side of a valley. The height of the ball up the valley side represents the energy density of the scalar field. When the ball is high up, the energy in the field is high, driving the inflationary expansion. As the ball rolls slowly down the valley the energy reduces and inflation turns off. At the valley floor, the ball oscillates back and

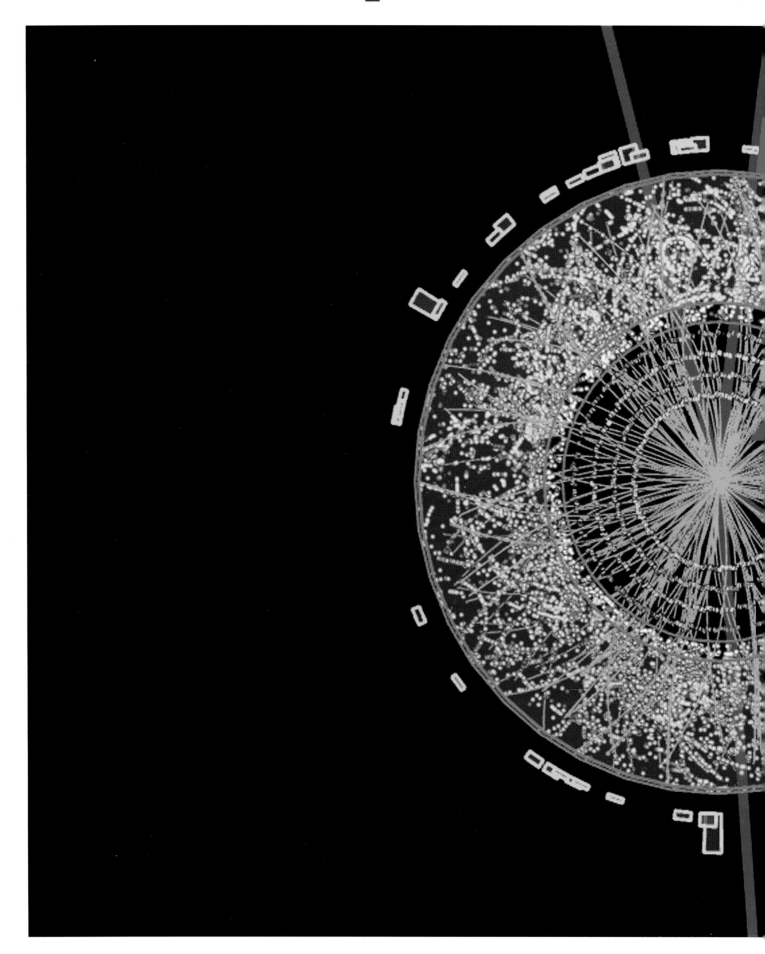

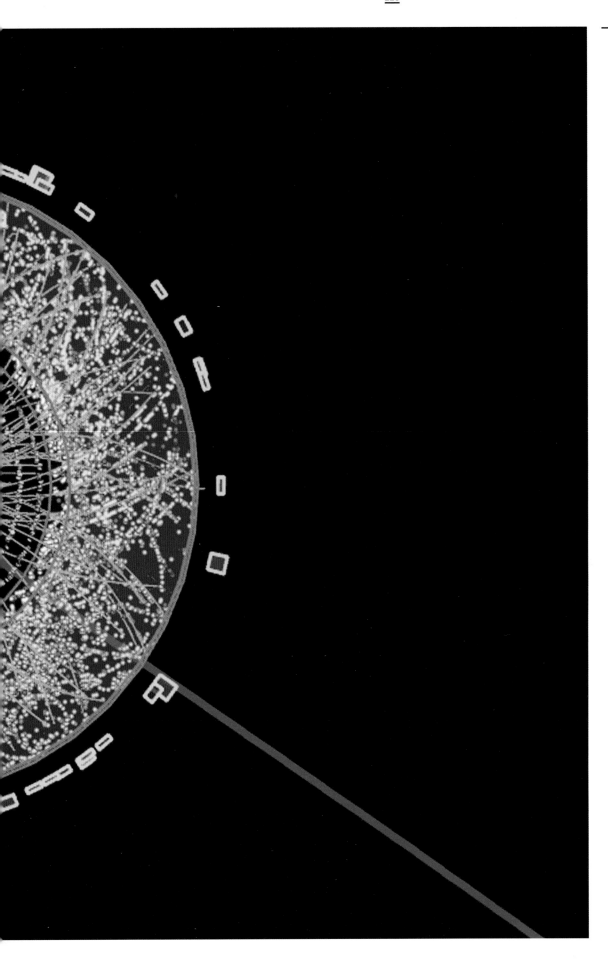

BICEP2 B-MODE SIGNAL

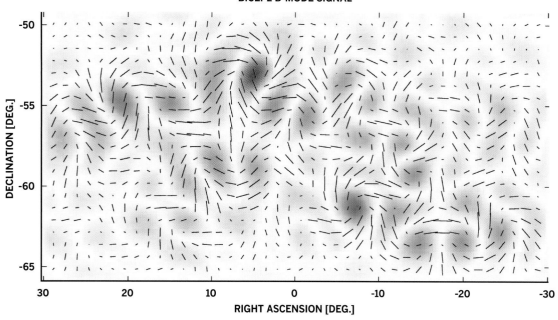

EVIDENCE FOR INFLATION
The results from the BICEP2 project in 2014 reveal the pattern detected in the cosmic microwave background which is a crucial piece of evidence to support the inflation theory in the study of the Big Bang.

forth until it comes to rest. The scalar field likewise oscillates and in so doing dumps its energy into the universe in the form of particles. In so doing it creates a hot dense soup, which we identify as the 'Big Bang'. In other words, inflation ends naturally and the standard Big Bang follows. The decay of the scalar field that drove inflation is the cause of the Big Bang!

Let us step back for a moment and recap with broad brush-strokes, because we seem to be wandering onto Leibniz's territory, and that's an astonishing place for physics to have arrived at. Our claim is that there exists a quantum field that causes the universe to expand exponentially fast for some period of time, and in doing so produces all the features of the universe we observe today, including the existence of galaxies and the matter out of which they are made. This is a triumph, and is now part of cosmology textbooks. Before the Big Bang, there was inflation. Fine, our philosopher friends would say, but what happened before inflation? Here, we must leave the textbooks and become a little more speculative, but not too speculative. We are still going to be working within the domain of mainstream physics.

There is an extension of what we might term standard inflationary theory. It is known as eternal inflation. Put simply, there seems to be no reason why inflation should stop everywhere at the same time. There should always be regions of the universe where the scalar field fluctuates to such high values that the exponential expansion continues, and these regions will always come to dominate the universe, however rare they may be, because they are exponentially expanding. Where inflation stops, Big Bangs herald the beginning of more sedately expanding regions like ours. But elsewhere, there is an ever-growing exponentially expanding universe, constantly spawning an infinity of Big Bangs. This theory, known as eternal inflation, leads to an infinite, immortal multiverse, growing fractal-like without end. This is truly mind-numbing, but we must emphasise that it is an entirely natural extension of standard inflationary cosmology.

Eternal inflation opens up even more exciting possibilities. As we discussed above, one of the great mysteries in physics today is the origin of the constants of nature such as the strength of gravity, the masses of the particles and the value of dark energy. These values appear to be fine-tuned for the existence of life, and understanding where they come from is a prerequisite for understanding our existence. In eternal inflationary models, each mini-universe can have different values of these constants and different effective laws of physics. The word 'effective' is important. The idea is that there is some overarching framework, out of which our laws and the constants of nature are selected randomly. If this is correct, then each of the infinite number of mini-universes that branch off the fractal inflationary multiverse can have different effective laws of physics, and all possible combinations will be realised somewhere. No matter how fine-tuned our laws appear for the existence of life, it is inevitable that such mini-universes as ours will exist, and there will be an infinite number of each possible set of combinations. There is no fine-tuning problem. Given the multiverse, we are inevitable. This is reminiscent of our rejection of your own personal uniqueness whilst listening to Joy Division at the beginning of the chapter. Yes, in isolation, the odds of you existing are almost vanishingly small. But given a mechanism for producing human beings, babies are born all the time and their existence is not surprising. Here, we have a mechanism for producing universes – and with an even greater statistical sledgehammer, the mechanism doesn't simply produce a few billion of them, it produces a potentially infinite number.

This is a quite stunning theoretical model, and I understand that it sounds like wild speculation. It isn't, though. Inflation is probably correct

ROSETTA PROBE
Our quest to understand the origins of our universe goes on and sees scientists breaking the boundaries of space travel in order to do so. On 6 August 2014 the European Space Agency's Rosetta spacecraft entered the orbit of the Churyumov-Gerasimenko comet, hoping to map and study the comet's surface and core to give us some more clues into how the solar system was formed some 4.6 billion years ago.

MULTIVERSE
We live in an infinite, eternal,
fractal multiverse comprised
of an infinite number of
universes like ours, alongside
an infinite number of universes
with different physical laws.
What does it mean if we exist
because it is inevitable? What
does it mean if the existence
of our universe is inevitable?
Perhaps we can only ask: what
does it mean to us?

INFLATION
Andrei Linde, 'Inflationary
Cosmology after Planck 2013',
arXiv:1402.0526v2 [hep-th].

in some form, in the sense that before what we call the Big Bang, there was an exponential expansion of spacetime. Scalar fields, which are known to exist, have the correct properties to drive such an expansion, although there are other theoretical models of inflation, as well. Theoretical physicists studying inflationary models have discovered that almost all of them are eternal, in the sense that they stop inflating in patches rather than all at once. This means that the potential for creating universes, in the guise of inflation, is always expanding faster than it is decaying away, and it will therefore never stop. We live in an infinite, eternal, fractal multiverse comprised of an infinite number of universes like ours, alongside an infinite number of universes with different physical laws. We exist because it is inevitable. Almost.

There is one very important caveat to this picture. Recent research suggests that eternal inflationary models may be eternal in the future, but not in the past. They never stop, but they may have to start. I can't give you a definitive answer to this ultimate question, because nobody yet knows. I can quote from Andrei Linde's recent review of inflationary cosmology, published in March 2014.

'In other words, there was a beginning for each part of the universe, and there will be an end for inflation at any particular point. But there will be no end for the evolution of the universe *as a whole* in the eternal inflation scenario, and at present we do not know whether there was a single beginning of the evolution of the universe as a whole at some moment t=0, which was traditionally associated with the Big bang.'

And so we reach the end. Defining the Big Bang as the initial hot, dense phase of our observable universe that gave rise to the CMB 380,000 years later, we understand what happened before. There was a period of inflationary expansion, which could have been driven by a scalar field in accord with the known laws of physics. That inflationary expansion is probably still going on somewhere, spawning an incalculable number of universes as we speak, and it will continue doing this forever. We live in an eternal universe, in which everything that can happen does happen. And we are one of the things that can happen. Did the whole universe have a beginning, an essential, external cause in the spirit of Leibniz's God? We still don't know. Possibly there was a 'mother of all Big Bangs', and if so, we will certainly need a quantum theory of gravity to say anything more.

What does this mean? The wonderful thing for me is that nobody knows, because the philosophical and indeed theological consequences of eternal inflation have not been widely debated and discussed. My hope is that in trying to summarise the issues, regrettably briefly and necessarily superficially in the television series and in a little more depth here, these ideas will be accessible to a wider audience and stimulate discussion. This is desirable and necessary, because ideas are the lifeblood of civilisation, and societies assimilate ideas and become comfortable with their implications through understanding and debate. If eternal inflation is the correct description of our universe, it will be the artists, philosophers, theologians, novelists and musicians, alongside the physicists, who explore its meaning. What does it mean if the existence of our universe is inevitable? What does it mean if we are not special in any way? What does it mean if our observable universe, with all its myriad galaxies and possibilities, is a vanishingly small leaf on an every-expanding fractal tree of universes? What does it mean if you are, because you have to be? I can't tell you. I can only ask – what does it mean to you?

For small creatures such as we, the vastness is bearable only through love.
Carl Sagan

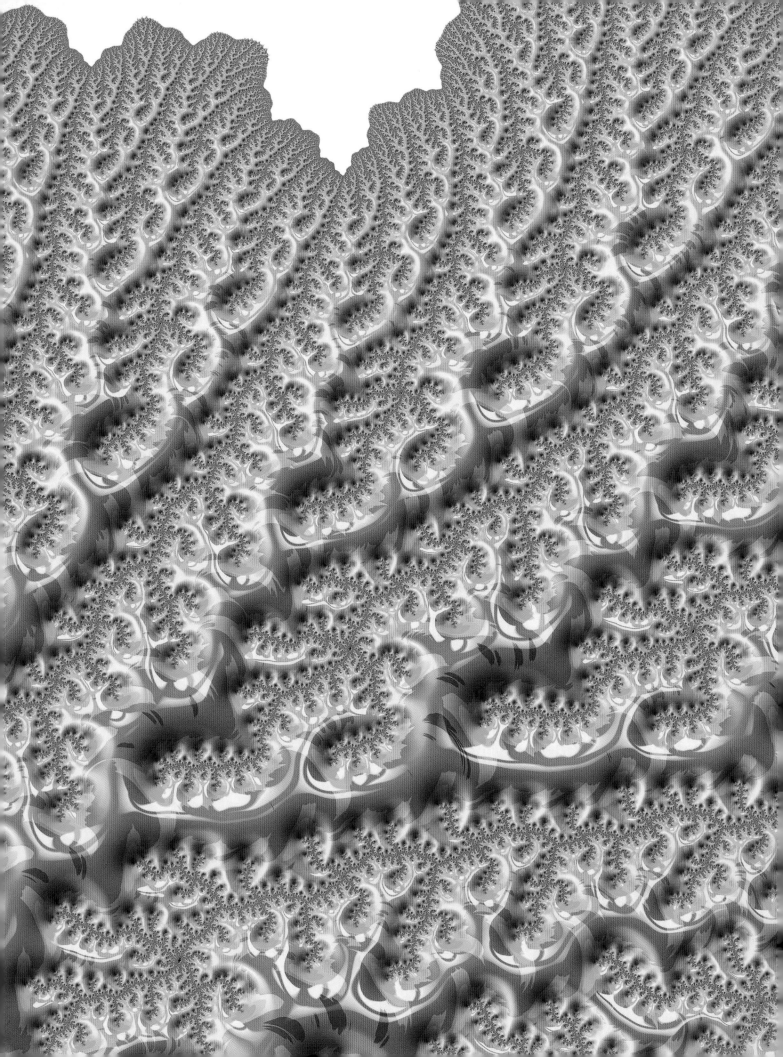

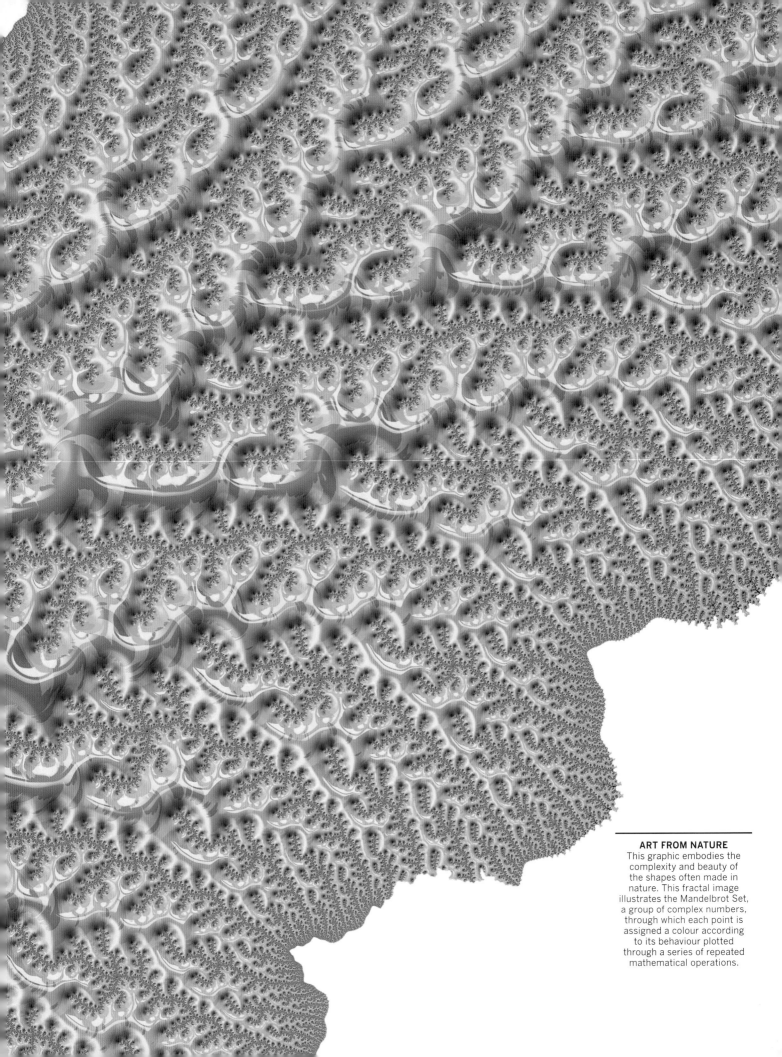

ART FROM NATURE
This graphic embodies the complexity and beauty of the shapes often made in nature. This fractal image illustrates the Mandelbrot Set, a group of complex numbers, through which each point is assigned a colour according to its behaviour plotted through a series of repeated mathematical operations.

WHAT IS OUR FUTURE?

I can hardly wait
To see you come of age
But I guess we'll both just have to be patient
'Cause it's a long way to go
A hard row to hoe
Yes it's a long way to go
But in the meantime
Before you cross the street
Take my hand
Life is what happens to you
While you're busy making other plans

John Lennon

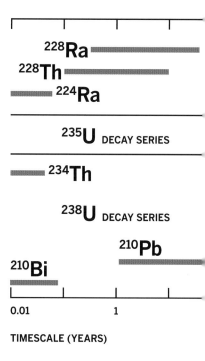

MAKING THE DARKNESS VISIBLE

Yet from those flames, no light; but rather darkness visible.
John Milton. Paradise Lost *1, 63.*

They must have descended into the darkness for a reason. Their burning dry-grass torches would have filled the caverns with acrid smoke, sucking the oxygen from the wet air. They would have moved carefully, fearfully perhaps, enveloped in a dim, flickering sphere of red, fading into a profound silent dark, the like of which I don't experience. A child held her hand against the rock, and blew a red-pigmented mixture across it with a straw. She smiled – 'my hand'. Her companions reached into the pigment and, in careful movements, inked a line of dots beside the handprint. The precision of a young imagination. A retreat to the lightness of the cave mouth. 'Perhaps we'll come back someday,' she thought.

Over 40,800 years later, I held my hand next to hers, because the experts on the Upper Paleolithic told me that the handprints are always those of children, and most likely always female. El Castillo in Northern Spain contains some of the oldest cave-art in the world. It is not known precisely how old, because the pigments themselves cannot be dated. The art is covered in calcite, which dripped and crystalised across the handprints and dots as the whole of recorded history played out above. Calcite contains uranium-234 atoms, which decay with a half-life of 245,000 years into thorium-230, which in turn decays with a half-life of 75,000 years. Thorium is not soluble in water, so there was none when the limestone formed. By measuring the concentrations of the uranium isotopes 234 and 238, and the thorium-230, a precise date for the formation of the calcite can be measured. This gives a minimum date for the art, since of course it must have been created before it was covered. The limestone covering the red dots formed 40,800 years ago. The oldest handprint was covered 37,300 years ago.

These dates are significant, because before 41,000 years ago there is no evidence of modern humans in Europe. Homo sapiens arrived tantalisingly close to the minimum age of the art in the darkness of El Castillo, leading some anthropologists to suggest that the art is not human. Rather, it may have been created by our close cousins, the Neanderthals, who dominated Europe at the time. I find this possibility profoundly interesting, and moving. It is interesting because the creators of this art had all the attributes that we might lazily refer to as 'uniquely human'. The retreat into the deep caves was undoubtedly a sophisticated

REACHING OUT FROM AN ANCIENT LIFE
Two tiny handprints give a tantalising glimpse into an ancient way of living in the Altamira caves in northwestern Spain.

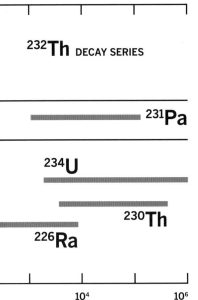

URANIUM SERIES DATING

Dating ranges of different nuclides within the three U-series decay chains to show their utility.

^{232}**Th** DECAY SERIES

^{231}**Pa**

^{234}**U**

^{230}**Th**

^{226}**Ra**

10^4 10^6

response to the world. This is not mere decoration, because cave-art like this is not found near the cave entrances where these 'people' lived. Its creation is highly ritualised. The darkness is integral. One of the most beautiful pieces in El Castillo is a bison, half-carved out of a column of rock and shaded with pigments to emphasise the arch of its back. When illuminated by torchlight, the rock casts a flickering, animal shadow onto the cave wall. The interaction of light and dark was important to the rituals carried out here before history, perhaps before humans. The cave resonates with ideas, curiosities and fears. It represents a border; the transition from existence to living. If this is a human place, it is a record of the first stumbling steps towards humanity. But if it is Neanderthal, it is a record of an ending, an ascent cut short. 'Perhaps we'll come back someday,' thought the little girl in my imagination. Not long afterwards, her species became extinct, out-competed by their incoming cousins. Perhaps. It is possible that the date coincides with the migration of Homo sapiens into Europe because the art is indeed human. Some anthropologists believe that the art may have been a response to the native Neanderthal population; a sort of prehistoric shock and awe, asserting cultural dominance and engendering a sense of community and superiority in the nascent human population. Things never change. If this is the case, the Neanderthals inadvertently played a role in our ascent. The roles may have been reversed, however. Perhaps our ancestors found a young, emerging and more sophisticated culture when they crossed the Mediterranean. A species distantly related to us whose desire to explore the darkness we assimilated. Perhaps our intellectual climb was, in part, a response to them. Intellectual superiority does not guarantee survival; witness the fall of classical civilisation.

This possibility is illustrative of a fact that we modern humans often subconsciously rest in the shadows. Things can end, for ever. Species become extinct, and that doesn't only apply to animals with feathers and no feelings. The Neanderthals became extinct, and they may have begun to imagine a future before they lost it. The red handprints of El Castillo are overwhelming in this context. Go there. Hold your hand up to hers, hear the giggles, picture the smiles, imagine the beginnings of hope, and listen to the silence.

At least 40,800 years later, we can use our knowledge of nuclear physics to move backwards through time to piece together her story. Science is a time machine, and it goes both ways. We are able to predict our future with increasing certainty. Our ability to act in response to these predictions will ultimately determine our fate. Science and reason make the darkness visible. I worry that lack of investment in science and a retreat from reason may prevent us from seeing further, or delay our reaction to what we see, making a meaningful response impossible. There are no simple fixes. Our civilisation is complex, our global political system is inadequate, our internal differences of opinion are deep-seated. I'd bet you think you're absolutely right about some things and virtually everyone else is an idiot. Climate Change? Europe? God? America? The Monarchy? Same-sex Marriage? Abortion? Big Business? Nationalism? The United Nations? The Bank Bailout? Tax Rates? Genetically Modified Crops? Eating Meat? Football? X Factor or Strictly? The way forward is to understand and accept that there are many opinions, but only one human civilisation, only one Nature, and only one science. The collective goal of ensuring that there is never less than one human civilisation must surely override our personal prejudices. At least we have come far enough in 40,800 years to be able to state the obvious, and this is a necessary first step.

'We've woken up at the wheel of the bus and
realised we don't know how to drive it'

MAKING THE DARKNESS VISIBLE

SUDDEN IMPACT

On 15 February 2013 at 9.13am a 12,000-tonne asteroid entered Earth's upper atmosphere travelling at 60 times the speed of sound. It came from the direction of the Sun, so there was never any chance of seeing its approach. The rock broke up at an altitude of 29 kilometres, depositing over twenty times the energy of the Hiroshima bomb into the sky above the Russian town of Chelyabinsk. Thousands of buildings were damaged by the shockwave and 1500 people were injured, mainly by flying glass as windows smashed in multiple cities across the region. Sound waves from the explosion rattled around the globe twice, and were detected by a nuclear weapons monitoring station in the Antarctic. The Russian parliament's foreign affairs committee chief Alexei Pushkov took to Twitter: 'Instead of fighting on Earth, people should be creating a joint system of asteroid defence.' Naive idealism? Over-reaction? Hollywood? Not really. Sixteen hours later, a 40,000-tonne asteroid named 367943 Duende streaked by at an altitude of 27,200 kilometres, well within the orbits of many of our satellites, although it missed them all. This one had a name, because it was discovered by astronomers in Spain in 2012. There is a 1 in 3000 chance that Duende will strike the Earth before 2069; if it does, it could destroy a city, which isn't too bad.

Before Chelyabinsk, the last recorded large impact was the Tunguska event over Siberia in 1908. The shockwave created by the airburst flattened 2000km² of forest in an energy release close to that of the United State's most powerful hydrogen bomb test at Bikini Atoll in March 1954. Events on this scale are thought to occur on average once every 300 years, and could easily wipe out a densely populated region. The best-known impact in popular culture was the Chicxulub event in Mexico's Yucatan Peninsula 66,038,000 ± 11,000 years ago, which wiped out the non-avian dinosaurs. Precision is important when available. If they'd had a space programme, Carl Sagan once quipped, or perhaps lamented, the dinosaurs would still be around, although in that case we wouldn't. The Chicxulub asteroid was probably around 9.5 kilometres in diameter, and the energy release of such an object exceeds that of the world's combined nuclear arsenal by a factor of a thousand. Or, if you like scary statistics, that's 8 billion Hiroshima bombs. Such events are estimated to occur on average every 100 million years, give or take, and are quite capable of

CAPTURING THE CRASH
These images show the meteor hurtling through the dark early morning sky at 9.20am just before it crashes into the mountains above Chelyabinsk.

TORINO SCALE

The Chicxulub impact, believed by many to be a significant factor in the extinction of the dinosaurs, has been estimated at 10^8 megatons, or Torino Scale 10. The impact which created the Barringer Crater and the Tunguska event in 1908 are both estimated to be in the 3–10 megaton range, corresponding to Torino Scale 8. The 2013 Chelyabinsk meteor had a total kinetic energy prior to impact of about 0.4 megatons, corresponding to Torino Scale 0. In all cases their impact probability was of course 1, as they actually hit Earth. As of May 2014, there are no known objects rated at a Torino Scale level greater than zero.

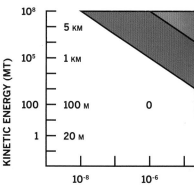

PROBABILITY OF IMPACT

destroying human civilisation and possibly causing our extinction. At the other end of the scale, rocks of around a millimetre in diameter hit the Earth at a rate of two a minute.

Alexei Pushkov was right. It is absolute idiocy not to pay attention to the danger of impacts from space, and fortunately our space agencies have begun to do so. NASA's Near Earth Object Program created the Sentry system in 2002, which maintains an automated risk table continually updated by new observations from astronomers around the world. I am writing these words on 3 September 2014, and there are currently no high-risk objects in the table, although there are 13 asteroids with the potential to impact Earth that have been observed within the last 60 days. The risk posed by an asteroid is quantified on the Torino Scale.

Every known near-Earth asteroid is assigned a value on the Torino Scale between 1 and 10, calculated by combining the collision probability with the energy of the collision in megatons of TNT (see table for 1–10 of the Torino Scale). Asteroid 99942 Apophis reached level 4 on the Torino Scale in December 2004. Initial observations and calculations suggested this 350-metre-wide asteroid had a 1 in 37 chance of a potential collision with the Earth on 13 April 2029 and a further chance of hitting us seven years later if it missed first time around. This would not have been a civilisation-threatening event, but it could have laid waste to a small country. Subsequent observations have effectively ruled out the risk from 99942 Apophis, but statistically speaking such an impact is expected to occur every 80,000 years or so. Although the Sentry table is currently benign, there are at least two very good reasons why we shouldn't relax and forget about impact risks. Firstly, we haven't detected all of the threatening objects by any means, as the Chelyabinsk event so effectively reminded us. And secondly, we don't currently know precisely what to do if we do observe an asteroid with our name on it, which could happen tomorrow. In 2015 a new early warning system called ATLAS (Asteroid Terrestrial Impact Last Alert Sytem) will come on line. Eight small telescopes will scan the sky for any sign of faint objects that may pose a threat to the Earth. ATLAS will give up to three weeks' warning of an impact, which is enough time to evacuate a large region, but probably not an entire country. The cost of our global insurance policy? One third of the annual wages of Manchester United striker Wayne Rooney. Such comparisons always sound childish of course; I'm well aware of how

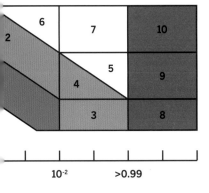

6	7	10
2	5	9
4		
3	8	

10^{-2} >0.99

FRAGMENTS FROM SPACE
A fragment from the meteor which caused so much damage as it hurtled through the skies to collide with Earth's surface.

MARS-500 PROJECT
The Mars-500 experiment was conducted between 2007 and 2011 at the Russian Institute of Biomedical Problems (IBMP) in Moscow, Russia. The final stage, held from June 2010 to November 2011, placed six crew members in an isolation facility to simulate a 520-day mission to Mars. The aim was to study the psychological and medical aspects of long-duration human spaceflights. Photographed during the 30-day 'Mars landing' period in February 2011, the photo shows Diego Urbana (Italy, ESA) carrying out a simulated 'Marswalk' using tools designed for Soviet lunar missions in the 1960s and 1970s.

capitalism functions, and I know that Wayne Rooney generates income for the Manchester United corporation in excess of his wages. But the aim of this chapter is to argue that there is a flaw in the majestic edifice of human civilisation; our myopic and cavalier disregard for our long-term safety. In my view, the reason for the short-sighted approach is that nothing catastrophically bad has happened to humanity in recorded history that we haven't inflicted upon ourselves, unless of course you believe in Noah's Ark, and even that was presumably down to us because one assumes that God is usually quite a patient sort. One of the central themes of this book has been to argue that the human race is worth saving because we are a rare and infinitely beautiful natural phenomenon. One of the other themes is that we are commonly and paradoxically ingenious and stupid in equal measure. I do not personally think that there is anyone out there to save us, and so it follows that we will have to save ourselves; at least, that would seem to me to be a good working assumption. This is why I don't feel naïve, idealistic or like a particularly radical member of the Student Union in a Che Guevara T-Shirt when I ask the question 'Is it reasonable to spend less on asteroid defence than on a footballer's annual salary?' When I look in the mirror and think about that, my face assumes an interesting shape – you should try it.

NASA is working hard in the face of apathy to do something to close the gap between the capabilities of the dinosaurs and us. Twenty metres beneath the surface of the Atlantic ocean, 8 kilometres off the coast of Key Largo, Florida, is the *Aquarius Reef Base*. Originally constructed as an underwater research habitat to study coral reefs, it is used by NASA to train astronauts for future long-duration space missions. The base allows for saturation diving, which greatly increases the length of time a researcher can spend exploring the reefs. On a normal scuba dive, a diver can spend a maximum of 80 minutes at a depth of 20 metres without having to go through decompression. The diver can remain at this pressure for several weeks, however, as long as they decompress when they return to the surface – a process that takes almost a day. Since the air pressure inside *Aquarius* is the same as the pressure outside in the water, researchers living inside the base can spend many hours a day exploring the sea bed using standard scuba equipment, but with the important caveat that they cannot return to the surface a few metres above their heads. If anything goes wrong, they must return to *Aquarius* and deal with the problem inside the base. For all practical purposes therefore, they are isolated; it's not possible to panic or simply loose patience and return to civilisation above. This is why NASA uses the *Aquarius* base to train astronauts to work in a hostile environment and test their psychological suitability for long-duration space missions.

Filming inside *Aquarius* was a personal highlight of *Human Universe*. We didn't want to have to decompress of course, so we had a strict time limit of 100 minutes inside the base spread over two dives. The ex-US Navy diver in charge of our dive was wonderfully clear as far as timings were concerned. 'If I say leave, you don't smile and take one more shot – you leave! Otherwise you stay, for a long time. Your choice. I know you media types.' *Aquarius* has the look and feel of a spacecraft from a science fiction film. There are six bunk beds piled three-high at one end, and a galley area complete with microwave and sink at the other. In between, there are control panels, some books on marine life, and a laptop computer station. Above the table, there is a single round window looking out across the reef. Through an air-lock-style exit, there is a dive platform with access to the scuba tanks and the open sea. NASA's Extreme Environment Mission Operations (NEEMO) team had just completed a nine-day mission when we arrived. Led by Akihiko Hoshide of the Japanese Aerospace Exploration agency, the mission was part of the long-term goal of landing astronauts on an asteroid, and developing the capability to deflect one, should the need arise. There are strong scientific and commercial reasons for exploring asteroids; they are pristine objects that will allow us to better understand the formation of our solar system over 4.5 billion years ago, and rich in precious metals precisely because they are pristine. On Earth, heavy metals such as palladium, rhodium and gold migrated into the Earth's core, leaving the accessible crust depleted. Asteroids are too small to have separated in this way, leaving the primordial abundances of these valuable metals untouched and accessible.

Whether for commercial, scientific or practical reasons, learning how to land on asteroids, exploit their resources and manipulate their orbits is clearly an eminently sensible thing to do. And make no mistake, we will have to move one at some point.

SEEING THE FUTURE

In the year 35,000 CE the red dwarf Ross 248 will approach the solar system at a minimum distance of 3.024 light years, making it the closest star to the Sun. 9000 years later it will have passed us by, ceding the title of nearest neighbour to Proxima Centauri once again. Coincidently, in 40,176 years, *Voyager 2* will pass Ross 248 at a distance of 1.76 light years. We know this because we can predict the future.

We've encountered Newton's laws several times in this book. In Chapter 3 we used them to calculate the velocity of the International Space Station in a circular orbit around the Earth. At a distance r from the centre of the Earth, the velocity v is

$$v = \sqrt{\frac{GM_e}{r}}$$

Let's look at this equation in a different way by rewriting it as

$$\frac{dx}{dt} = \sqrt{\frac{GM_e}{r}}$$

Here, we've used the notation of calculus. That may strike fear into your heart if you haven't done any mathematics since school, but don't worry. All we need to know is the meaning of the symbol

$$\frac{dx}{dt}$$

In words, this denotes the rate of change of the position of the space station with respect to time, otherwise known as its velocity v. You have an intuitive feel for this even if you've never done any mathematics. If you get into your car and drive it away from your house in a straight line at a velocity of 30 kilometres per hour, then in one hour you will be at a position 30 kilometres away from your house in the direction in which you drove the car. The equation is telling us what the position of the Space Station *will be* at some point later in time, given knowledge of where it is and how it is moving in the present. It predicts the future. This sort of equation is known as a *differential equation*. In Chapter 4 we wrote down the 'rules of the game' – Einstein's General Theory of Relativity and the Standard Model of particle physics. The notation is a little more complicated, but in the Standard Model you'll notice the symbols D_μ and δ_μ, which are more complicated versions of

$$\frac{dx}{dt}$$

In Einstein's equations, there are also these so-called derivatives hidden away in the compact mathematical notation. The known fundamental laws of physics all function in this way. Given knowledge of how some system or collection of natural objects is behaving *now*, we can compute what they will be doing at some time in the future. The system in question may be a solar system, a collection of atoms and molecules, or the weather. There are practical limitations, of course, and the weather forecast is a good example. Earth's climate system is very complicated, with many hundreds of thousands of variables. Ocean currents in the Pacific might affect future rainfall in Oldham, and so long-term forecasting of local weather conditions comes with increased uncertainty.

People do of course make statements, often based on human experience rather than science, which are more likely to be right than wrong. Red sky at night, Shepherd's delight. Red sky in the morning, Shepherd's warning.

OUR NEIGHBOURHOOD
A 3-D diagram of the stars nearest to our solar system.

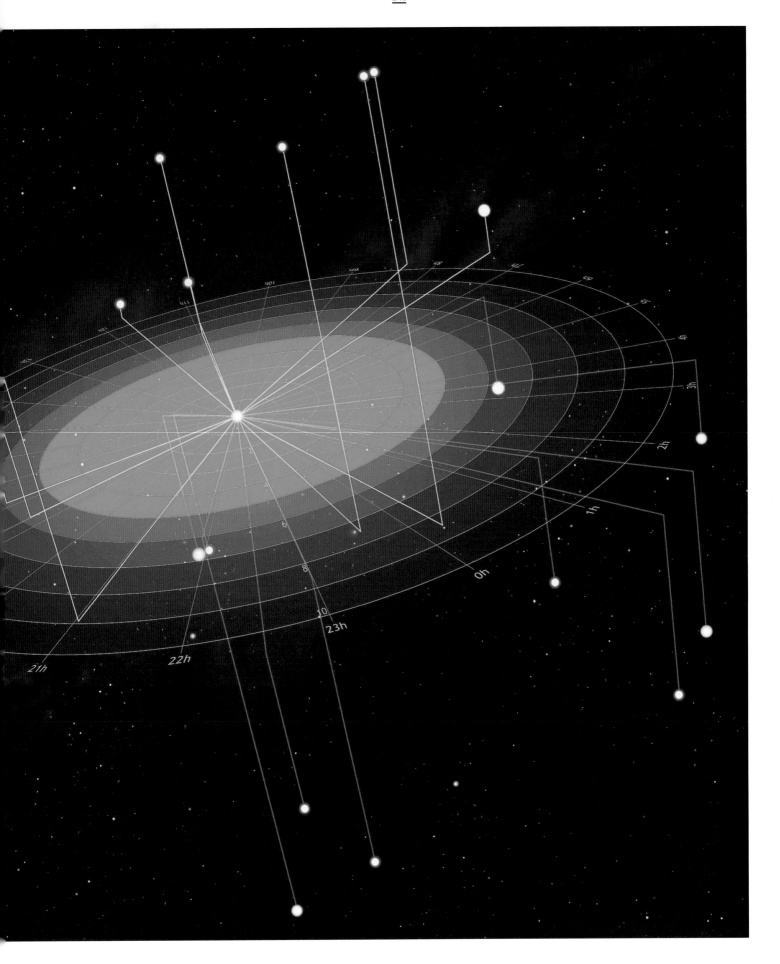

RED SKY AT NIGHT
The laws of physics can be used to describe weather patterns and the motion of the stars.

This is often true in countries like the UK whose weather is dominated by westerly winds, because a red sunset is usually a sign of high pressure to the west, which is associated with fine weather. But if you're doing well in a statistically significant sense using 'folklore' or 'ancient wisdom', it's because the patterns and regularities you are using to make your predictions emerge from underlying physical laws, which are described by differential equations. The laws of physics in essence reflect the underlying simplicity of nature and the regularity with which it behaves. They are not magic. We can describe the natural world using mathematics *because* it is regular and behaves consistently. It is my opinion that we *must* observe a universe that behaves in a regular and consistent way because such behaviour is necessary for complex structures like brains to evolve. A universe of anarchy, with sub-atomic particles interacting without some sort of framework or rules, would surely not support life, or indeed any structures at all. This is known as a selection effect. We observe a universe whose behaviour can be described by a limited set of

differential equations because we wouldn't exist if it were not so. This is my opinion, and there are scientists and philosophers who might disagree. It could be the case that there is no simple underlying framework to the universe, and our success to date has deceived us. Or perhaps the ultimate laws are and will forever remain beyond human understanding. We might simply not be smart enough to figure them out. There are also systems that cannot be described using differential equations. The patterns generated in Conway's Game of Life are an example, where algorithmic rules are used to generate complex patterns and even computing devices such as Turing machines. But what can be said with certainty is that, as far as we can tell, the natural world does behave in a way that is amenable to a description based on the differential equations of physics, and these allow us to predict the future, given knowledge of the present. This is why our asteroid defence system will work if we make enough high-precision observations of the sky. Sort of.

Ahhh, caveats. There are always caveats.

213.4kg

$$\Delta v = v_e \ln \frac{m_0}{m_1}$$

$$E_k = \frac{1}{2}mv^2$$

SCIENCE VS. MAGIC

Chaos: When the present determines the future, but the approximate present does not approximately determine the future.
Edward Lorenz

We should be confident in science. It works. But it has limitations, some of which are fundamental. We've encountered Newton's laws of motion and gravitation time and again in this book. They are very simple – the archetypal physical laws – and are used every day by engineers, navigators and asteroid watchers. One of the simplest imaginable real-life systems to which Newton's law of gravitation can be applied is a single planet orbiting around a single star. For this case, Newton's laws allow for a precise prediction of the future position of the planet. The orbit is predictable and periodic, which is to say the planet returns to precisely the same position around the star every orbit. It's clockwork – the way the solar system is often pictured. If a third object – a moon, say – is introduced, it was proved in the late nineteenth century by Heinrich Burns and, later, Henri Poincaré, that no general solution to Newton's equations can be found. There are a handful of special cases, which are still being discovered, for which there are repeating solutions, but in general, the orbits of three bodies acting under gravity never repeat; their motion around each other traces out a tremendous ever-changing mess! This isn't a failure of mathematics. Natural systems really do behave in this way. The solar system is a case in point. The planets orbit like clockwork on timescales of millions of years, but we are currently unable to predict the Earth's orbit for more than 60 million years into the future. Beyond

THE POWER OF THE SUN
The Sun is a powerful marker of the seasons across ancient civilisations such as the Mayans in Chichen Itza, Mexico.

NATURE'S TIMEPIECE
The alignment of the Sun through carefully positioned stones served as an effective guide to the seasons for our ancient ancestors.

that, the sensitivity of the predictions to uncertainties in our current knowledge of the Earth's orbit, and the gravitational influence of other bodies in the solar system, become too great. This isn't only a reflection of our lack of knowledge. It also reflects an important fundamental point, which is that solar systems such as ours *are* unstable over long timescales. Their behaviour is chaotic; the apparent clockwork can break down into a whirling unpredictable swarm. Recent simulations suggest that Mercury *could* be wrenched out of its orbit and collide with the Sun, and that even the Earth *may* have a close encounter with Venus or Mars on time periods of 3–5 billion years. The words *could* and *may* are italicised for a reason. These predictions are statistical in nature – it is estimated that there is a 1 per cent chance that Mercury will be thrown into a much more elliptical orbit during the next 5 billion years. The uncertainty is down to the extreme sensitivity of the predictions to what physicists call the initial conditions – the current knowledge of precisely where everything in the solar system is at this instant, and how everything is moving at this instant. Other errors are caused by our precise knowledge of the mass and shape of all the objects in the solar system, not to mention the slight perturbations from incoming comets and the ever-shifting asteroids. The area of physics and mathematics concerned with such systems is called chaos theory, and as the pioneer of the field Edward Norton Lorenz put it, nature's complexity usually leads to a situation in which approximate knowledge of the present, which is in practice all we ever have, does not approximately determine the future.

For the asteroid hunters, this is an intensely problematic truth. It is not possible to observe an asteroid once, and then pop its position and velocity into a computer to work out whether it will ever hit us. Instead, a gravitational keyhole system is used. A keyhole is a small volume of

STONEHENGE
One of the earliest examples of stone markers used to determine the date of the summer solstice, Stonehenge dates back to 3000 BCE.

EQUINOXES AND SOLSTICES

When the Sun is crossing the celestial equator, day and night are of nearly equal length at all latitudes, which is why these dates are called the equinoxes ('equal nights'). In March, as the Sun is moving northwards along the ecliptic, this is called the vernal equinox, and in September as the Sun is moving southwards we refer to it as the autumnal equinox. The times when the Sun is at its furthest from the celestial equator are called the summer and winter solstices. The world 'solstice' comes from the Latin meaning 'Sun stands still' because the apparent movement of the Sun's path north or south stops before changing direction.

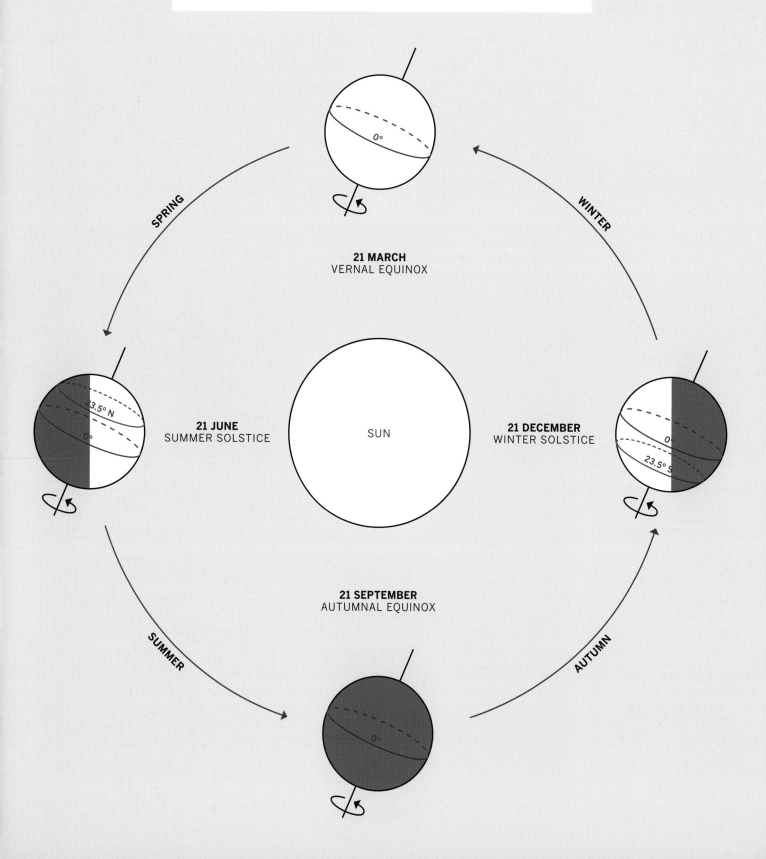

SPRING

WINTER

21 MARCH
VERNAL EQUINOX

23.5° N

0°

21 JUNE
SUMMER SOLSTICE

SUN

0°

23.5° S

21 DECEMBER
WINTER SOLSTICE

21 SEPTEMBER
AUTUMNAL EQUINOX

SUMMER

AUTUMN

0°

Philos. Trans. Vol. XLVIII TAB XIX *p. 520.*

T. Ferguson inv.et del.   *J. Mynde Sculp.*

PREDICTING ECLIPSES
An eighteenth-century solar eclipse predictor, based on observations of the natural world.

space close to the asteroid's current orbit. If an asteroid passes through the keyhole, perhaps because of a gravitational nudge from some other object in the solar system, then it is highly likely that it will impact the Earth on its next pass. 99942 Apophis was assigned such a keyhole in 2004 when it was classified at 4 on the Torino Scale. Fortunately, it didn't pass through, and this is why it is currently classified as harmless. The keyhole system reflects the fundamental unpredictability of complex physical systems over long timescales. This is why we have to keep observing and retain a keen understanding of the fundamental limits of our calculational prowess. Science isn't magic. This realisation is of course important in a practical sense if one is interested in saving the planet from asteroid impact. But it is also very important to embed caution and humility into our/my polemical celebration of the power of science. Scientific predictions are *not* perfect. Scientific theories are *never* correct. Scientific results are always preliminary. Whole fields of study can be rendered obsolete by new discoveries. But, I insist, science is the best we can do because it is not simply another arbitrary system of thought based on dreamt-up human axioms. It is the systematic study of nature, based on observations of the natural world and our understanding of those observations. Scientific predictions are not matters of opinion. At any given time, science provides the best possible estimate of what the future might bring, given our current understanding. The predictions may be wrong, they may be inaccurate, the errors may be fundamental in origin, but there is simply no other rational choice than to act according to the best available science, imperfect though, by necessity, its predictions will always be.

THE WONDER OF IT ALL

As of September 2014, in a population of 7.24 billion, 545 people have been to space, 24 people have broken free of the Earth's gravitational pull and 12 have landed on another world.

In 2013 Charlie and Dorothy Duke, a retired, church-going couple from New Braunfels, Texas, reached their 50th wedding anniversary. With two grown sons and nine grandchildren, Charlie and Dottie must have celebrated a life well lived, captured in photographs adorning the walls and mantelpieces of the family home. There is one Duke family photograph, however, that holds a unique place in history. I myself have a copy of it on my wall at home, signed by Charlie, and it's one of my favourite things. The photograph, taken in 1972, is an image of Charlie, Dottie and their two young sons Charles and Thomas when they were just six and four years old. The picture itself is of no particular note – a simple portrait of a family in 70s clothes, sitting on a bench in a garden. It's not dissimilar to the one below, which is me and my grandad photographed at around the same time. I was in Oldham, the Dukes were in Florida.

The reason I have a copy of the Dukes' photograph is not what it is – we are not related – but where it is. Charlie and Dorothy Duke are the only grandparents on Earth who can point their grandchildren's eyes towards the Moon and tell them there is a photo of Grandma, Grandpa, Dad and Uncle resting on the surface.

Charlie Duke was the Pilot of *Orion*, the Apollo 16 Lunar Module. At the age of thirty-six he remains the youngest human ever to have walked on the Moon. Together with Commander John Young, my childhood hero, the two astronauts spent three days in late April 1972 exploring the Descartes Highlands, covering almost 27 kilometres in the Lunar Rover.

The primary scientific aim of the mission was to explore the geology of the lunar highlands. It was thought that the unique rock formations around the landing site were formed by ancient lunar volcanism, but Young and Duke's exploration demonstrated that this explanation was incorrect. Instead, the landscape had been forged by impact events, scattering material outwards from the craters and littering the surface with glass. After three days on the lunar surface and setting a lunar land-speed record of 17km/h, Charlie Duke removed his family portrait from his spacesuit pocket, placed it on the lunar surface and snapped it with his Hasselblad. Inscribed on the back are the words 'This is the family of Astronaut Duke from Planet Earth. Landed on the Moon, April 1972.'

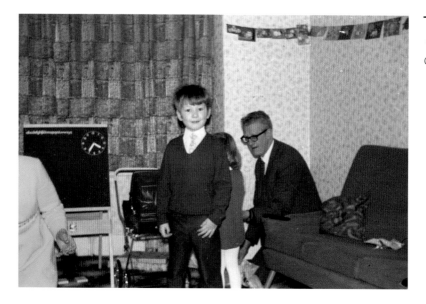

OLDHAM
A photo of my grandfather and me, taken at the same time as Charlie Duke left his own family photo on the Moon.

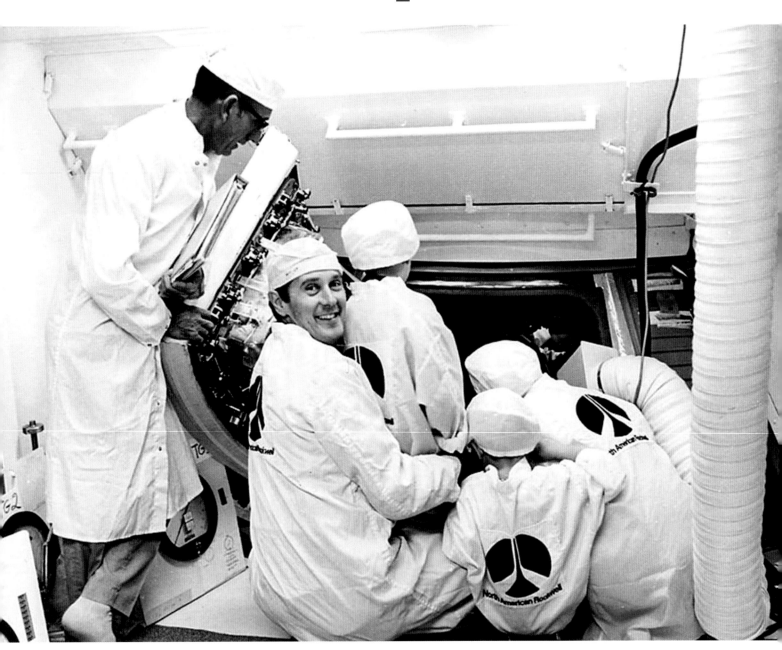

I remember being four years old in Oldham when Apollo 16 was on the Moon. Forty-two years later I talked to Duke for hours in a diner in Texas, with absolutely no regard at all for the film crew trying to make *Human Universe*. 'When I stepped onto the Moon it occurred to me that nobody had ever been here before. You looked out onto the most pristine desert – the most incredible beautiful place I've ever seen. No life, nothing like Earth, the rolling grey lunar surface with the blackness of space above.'

How ambitious was Apollo, I asked? 'They gave us eight and a half years to do it and we did it in eight years and two months. Nobody even knew how to do it,' replied the test pilot, who was used to doing things that nobody can do. 'Yeah sure. 15 minutes in space and we're going to land on the Moon in eight and a half years? But the remarkable part is that we did it, and I had a part in it.' Would it be possible now? 'No. We don't have the manpower to do it. Four hundred thousand people and unlimited budget and you can do a lot, and that's what we had!' What do you say to people who criticise manned exploration? There is surely more to human exploration than just science. 'It's the wonder of it all,' replied the astronaut. 'And that's what we bring – what manned flight brings to

FIRST FAMILY ON THE MOON
Charlie Duke (pictured above) made sure his family shared in his adventure by leaving a photo of them on the lunar surface.

the human spirit, the human being – the wonder of it all. The beauty of the universe, the orderliness of the universe, and you see it with your own eyes and it just captures your imagination. Let's see it, let's do it and let's discover it – that's been the human spirit all along.'

I think Apollo is the greatest human achievement. People argue with me of course. Gil Scott-Heron wrote a song called 'Whitey's on the Moon'. 'A rat done bit my sister Nell, with Whitey on the moon. Her face and arms began to swell, and Whitey's on the moon. I can't pay no doctor bill, but Whitey's on the moon. Ten years from now I'll be payin' still, while Whitey's on the moon.' The economics of Apollo are interesting. As Charlie said, the budget was whatever it had to be to get to the Moon by 1970. At the peak of spending in 1966, NASA received 4.41 per cent of the Federal budget, equivalent to around $40 billion today. That's a lot of money – almost half of the United Kingdom's annual debt interest bill. That's meant to be sarcastic, of course. The total cost of Apollo was in the region of $200 billion at today's prices, which is around a quarter of the cost of the UK's bank bailout programme initiated in October 2008. That's unfair, a City-type might splutter over a glass of Dom Ruinart, because that money was an investment in financial stability and has been repaid, give or take the odd £100 billion, which is neither here nor there. My reply would be yes, but Apollo was probably the most savvy investment in modern history. In 1989, the then US President George Bush said Apollo provided 'the best return on investment since Leonardo da Vinci bought himself a sketchpad'. Many academic studies have been carried out, and the most commonly quoted figure is that for every $1 spent on Apollo, $7 was returned to the economy over the period of a decade. Why? Because Project Apollo was conceived and executed in a tremendously smart way, distributing high-technology jobs and R&D projects across the country. It was also unarguably inspirational, propelling thousands of kids into science and engineering. The average age in Mission Control, Houston, on 20 July 1969 when Neil Armstrong landed on the Moon was 26. The old man in charge, Gene Kranz, was 36, and the old man flying the lunar module was 35. What happened to all those brilliant engineers? They went out into the economy of course, took the technology and expertise developed for the Moon landings and invented the modern world. The kids they inspired became known as Apollo's Children; the generation of optimists steeped in possibility who powered the United States economy through the last third of the twentieth century. The world loves this America, the one that flies to the Moon not because it's easy but because it's hard. I think America has lost its way, which might seem rich from a citizen of a small island that spends more on the wages of Premier League footballers annually than it does on research into the physical sciences and engineering, including its contributions to CERN, the European Space Agency and all UK-based scientific facilities. We've lost our way too, and so has the world. The World Bank defines R&D as 'current and capital expenditures (both public and private) on creative work undertaken systematically to increase knowledge, including knowledge of humanity, culture, and society, and the use of knowledge for new applications'. The United States spent 2.79 per cent of its GDP on increasing knowledge in 2012 – the UK spent 1.72 per cent. It has been estimated that the return on R&D spending in today's world economy is approximately 40:1. Imagine what we could do if we took these figures seriously.

My grandad, sitting behind me in that 1972 family Christmas photograph, was born in 1900. He was three years old when, on 17 December 1903 at Kill Devil Hills in North Carolina, Orville Wright took the controls of the Wright Flyer and lifted off the ground for twelve seconds. He was 68 when he saw Neil Armstrong walk on the Moon. Orville Wright himself died in the year that Neil Armstrong began studying aeronautical engineering at Purdue University, Indiana. I still find it hard to believe that I have spoken to someone who was born before

WALKING ON THE MOON
Charlie Duke is shown collecting lunar samples at Station no. 1 during the first Apollo 16 extravehicular activity at the Descartes landing site. This photograph was taken by Astronaut John W. Young, commander. The parked Lunar Roving Vehicle can be seen in the left background.

the human spirit, the human being – the wonder of it all. The beauty of the universe, the orderliness of the universe, and you see it with your own eyes and it just captures your imagination. Let's see it, let's do it and let's discover it – that's been the human spirit all along.'

I think Apollo is the greatest human achievement. People argue with me of course. Gil Scott-Heron wrote a song called 'Whitey's on the Moon'. 'A rat done bit my sister Nell, with Whitey on the moon. Her face and arms began to swell, and Whitey's on the moon. I can't pay no doctor bill, but Whitey's on the moon. Ten years from now I'll be payin' still, while Whitey's on the moon.' The economics of Apollo are interesting. As Charlie said, the budget was whatever it had to be to get to the Moon by 1970. At the peak of spending in 1966, NASA received 4.41 per cent of the Federal budget, equivalent to around $40 billion today. That's a lot of money – almost half of the United Kingdom's annual debt interest bill. That's meant to be sarcastic, of course. The total cost of Apollo was in the region of $200 billion at today's prices, which is around a quarter of the cost of the UK's bank bailout programme initiated in October 2008. That's unfair, a City-type might splutter over a glass of Dom Ruinart, because that money was an investment in financial stability and has been repaid, give or take the odd £100 billion, which is neither here nor there. My reply would be yes, but Apollo was probably the most savvy investment in modern history. In 1989, the then US President George Bush said Apollo provided 'the best return on investment since Leonardo da Vinci bought himself a sketchpad'. Many academic studies have been carried out, and the most commonly quoted figure is that for every $1 spent on Apollo, $7 was returned to the economy over the period of a decade. Why? Because Project Apollo was conceived and executed in a tremendously smart way, distributing high-technology jobs and R&D projects across the country. It was also unarguably inspirational, propelling thousands of kids into science and engineering. The average age in Mission Control, Houston, on 20 July 1969 when Neil Armstrong landed on the Moon was 26. The old man in charge, Gene Kranz, was 36, and the old man flying the lunar module was 35. What happened to all those brilliant engineers? They went out into the economy of course, took the technology and expertise developed for the Moon landings and invented the modern world. The kids they inspired became known as Apollo's Children; the generation of optimists steeped in possibility who powered the United States economy through the last third of the twentieth century. The world loves this America, the one that flies to the Moon not because it's easy but because it's hard. I think America has lost its way, which might seem rich from a citizen of a small island that spends more on the wages of Premier League footballers annually than it does on research into the physical sciences and engineering, including its contributions to CERN, the European Space Agency and all UK-based scientific facilities. We've lost our way too, and so has the world. The World Bank defines R&D as 'current and capital expenditures (both public and private) on creative work undertaken systematically to increase knowledge, including knowledge of humanity, culture, and society, and the use of knowledge for new applications'. The United States spent 2.79 per cent of its GDP on increasing knowledge in 2012 – the UK spent 1.72 per cent. It has been estimated that the return on R&D spending in today's world economy is approximately 40:1. Imagine what we could do if we took these figures seriously.

My grandad, sitting behind me in that 1972 family Christmas photograph, was born in 1900. He was three years old when, on 17 December 1903 at Kill Devil Hills in North Carolina, Orville Wright took the controls of the Wright Flyer and lifted off the ground for twelve seconds. He was 68 when he saw Neil Armstrong walk on the Moon. Orville Wright himself died in the year that Neil Armstrong began studying aeronautical engineering at Purdue University, Indiana. I still find it hard to believe that I have spoken to someone who was born before

39

MAN'S LAST MOONWALK
The crew of Apollo 17 take man's last steps on the lunar surface in 1972. The eyes of the world are now trained on Mars.

MAN ON MARS?
We've had the man on the Moon. Now man's mission is to travel 225 million kilometres to the inhospitable planet of Mars.

powered flight, and to someone who walked on the Moon. It is important to notice that this sentence can't be followed. Someone who walked on the Moon, comma, and someone who .... What? Where will the next generation of Apollo's Children come from? Perhaps a new superpower will take America's place as the great exploring nation. China and India, those re-emergent cradles of civilisation, have ambitions in space. As Jacob Bronowski wrote in the *Ascent of Man*, 'Humanity has a right to change its colour.' But I share his regret that the retreat of Western civilisation may leave Shakespeare and Newton as historical fossils, in the way that Homer and Euclid are. If that is the case, it will be our choice.

Two more astronauts followed Duke and Young onto the lunar surface. They left at 10.55pm GMT on 14 December 1972. Commander Gene Cernan, as he prepared to step on to the ladder of the Lunar Module, quietly spoke the final words from the Moon.

... I'm on the surface; and, as I take man's last step from the surface, back home for some time to come – but we believe not too long into the future – I'd like to just say what I believe history will record. That America's challenge of today has forged man's destiny of tomorrow. And, as we leave the Moon at Taurus-Littrow, we leave as we came and, God willing, as we shall return, with peace and hope for all mankind. Godspeed the crew of Apollo 17.
Gene Cernan, Taurus-Littrow Valley, 14 December, 1972.

SATURN V
The Apollo 11 Saturn V space vehicle lifted off with astronauts Neil A. Armstrong, Michael Collins and Edwin E. Buzz Aldrin Jr. at 9.32am on 16 July 1969 from Kennedy Space Center's Launch Complex 39A.

DREAMERS, PART 1

Apollo was about many things. It was about winning a race against the Soviets. It was about national pride. It was born out of fear as well as optimism. It was about laying the foundations of American dominance in the late twentieth century. It was about economic stimulus. It was about dreams. It succeeded on all fronts. Was it really about dreams? 'Well, space is there, and we're going to climb it, and the Moon and the planets are there, and new hopes for knowledge and peace are there. And, therefore, as we set sail we ask God's blessing on the most hazardous and dangerous and greatest adventure on which man has ever embarked.' I think so. Kennedy was a politician, but I believe he meant it.

So what of the dreamers now? Is the twenty-first century the era of pragmatism? The era in which we believe, because we have to, that the interests of shareholders are aligned with the interests of humanity? Innovation funds the shops on New Bond Street, but is that all? A common governmental lament is that new knowledge is not converted efficiently enough into economic growth. Is that what knowledge is for? Who pays for progress? Who *should* pay for progress?

Human Universe is a piece of documentary television, and this book is based on the series. Television is about stories; examples that illustrate a point. *Human Universe* is also at heart optimistic, because I am optimistic. I think we as a civilisation could do better, as I'm sure you've gathered, but it would be ridiculous to suggest that we are not doing some things right. In the final episode, we found two stories that demonstrate that long-term thinking is not dead; one almost Apollo-like in state-funded grandeur, and the other more modest but equally important. The first was a project I'd visited once before, back in 2009, known as the National Ignition Facility at the Lawrence Livermore National Laboratory in California. The aim is to make a star on Earth.

Nuclear fusion is the power source of the stars. The Sun releases energy in its core by turning hydrogen into helium. Two protons approach each other at high speed, because the core is hot. The core became hot initially through the collapse of the gas cloud which formed the Sun. Protons are positively charged, and therefore repel each other through the action of the electromagnetic force, but if they get close enough, the more powerful nuclear forces take over. The weak nuclear force acts to turn the proton into a neutron, with the emission of a positron and an electron neutrino. The proton and neutron then bind together under the action of the strong nuclear force to form a deuterium nucleus, which is an isotope of hydrogen (because it contains a single proton) with a neutron attached. Very quickly, another proton fuses with the deuteron to form helium-3, and finally two helium-3 nuclei stick together to form helium-4, with the emission of the two 'spare' protons. The important result in this convoluted process is that four protons end up getting converted into a single helium-4 nucleus, made of two protons and two neutrons, and the helium-4 nucleus is less massive than four free protons. This missing mass is released as energy, in accord with Einstein's equation $E=mc^2$, and this is why the Sun shines. The energy released in fusion reactions is colossal by terrestrial standards. If all the protons in a cubic centimetre of the solar core were to fuse into deuterium, enough energy would be produced to power the average town for a year. Or to put it another way, one kilogram of fusion fuel produces as much energy as 10 million kilograms of fossil fuel, which is approximately a hundred thousand barrels of oil, with no CO_2 emissions; the waste product is helium, which can be used to fill party balloons.

Energy is the foundation of civilisation. Access to energy underpins everything, from public health to prosperity. Access to clean water is

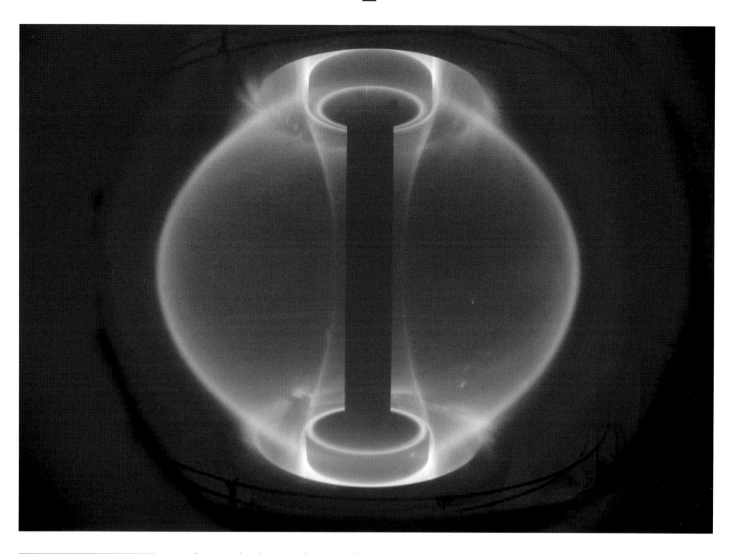

NUCLEAR FUSION RESEARCH
Spherical ball of plasma inside
the START device at Culham,
Oxfordshire, England.

surely more fundamental, you might say, but this requires energy. Even in the most arid regions, desalination plants or deep wells can deliver water in abundance *if* sufficient energy is available. It isn't, of course. Profligate energy use has a bad name today, but consider this. In every country in which the per capita energy use is greater than half the European average, adult life expectancy is greater than 70 years, literacy rates are greater than 90 per cent, infant mortality rates are low and more than one in five of the population is in higher education. The reason energy use has a bad name is not because it is bad in itself. It is good, it is the foundation of modern civilisation, and modern civilisation is a good thing. I don't want to live on a subsistence farm, sleep in stifling heat, run the risk of dying of malaria and have no access to clean water or cutting-edge medical care. I am lucky. I live in a city, I buy all the food I want from nice shops, I have a fulfilling job in a university and I get to do research at places like CERN, which is interesting. I want everyone in the world to have choices, like I have, and that means I want everyone in the world to have access to energy, like I have. In 2011, 1.3 billion people lacked access to electricity. Yes. Energy use is good. The problem with energy is how we produce it.

The world produces more than 80 per cent of its energy by burning fossil fuels. This is expected to fall to 76 per cent by 2035 as nuclear and renewables grow in importance. Burning things is humanity's oldest technology. The energy sector is responsible for two-thirds of global greenhouse gas emissions. The most recent scientific modelling suggests that global average temperatures will rise by around 2–2.5°C above the average of the years 1986 to 2005 by 2100. The rise could be less – as low

CONTAINING THE SUN
The inside of the target chamber at the National Ignition Facility (NIF), at the Lawrence Livermore National Laboratory, California, USA. This site is attempting to initiate and control hydrogen fusion as a sustainable energy source for future use. The beams from 192 lasers are focused here onto a 2-millimetre-wide capsule of deuterium–tritium (DT) gas. The total energy focused is 1.8 megajoules.

GLOBAL WARMING
The most recent summary is contained in the IPCC Climate Change 2014 Summary for Policy Makers: http://ipcc-wg2.gov/.

as 1 to 1.5°C, or it could be 4°C or more. Some of the uncertainty depends on our actions, and so there are assumptions about future behaviour built into the predictions. But over 90 per cent of computer models agree that global temperatures will have increased by 2100 as a result of greenhouse gas emissions from fossil fuel burning.

Nuclear fusion, then, is a good idea. If it can be made to work in an economically viable way, it will provide limitless, clean energy for everyone. It is not the *only* way of achieving this goal. One can make a case for solar power, and indeed an increased contribution from other renewables and nuclear fission. But it is a possible way to solve the world's energy problems for good, in principle, and is therefore worth exploring.

The challenge is technical rather than fundamental, in the sense that we know fusion works because the Sun does it. Fusion is difficult to achieve on Earth primarily because of the colossally high temperatures and pressures required. There are two approaches being followed, and each is Apollo-like. In Europe, a worldwide collaboration involving Russia, USA, the European Union, Japan, China, Korea and India is in the process of constructing ITER. This machine is in effect a magnetic bottle, which can store a plasma at temperatures in excess of 150 million °C – ten times that of the solar core. ITER will use deuterium and tritium, which is another isotope of hydrogen comprising one proton and two neutrons, to make helium-4. This bypasses the slow initial weak interaction in the Sun that makes deuterium out of hydrogen, making ITER a lot more efficient than our star. Deuterium is extracted from seawater, and tritium is made inside the reactor itself by irradiating a lithium blanket with the spare neutrons produced during the fusion reaction. An 800 MW fusion power station of this type would consume around 300 grams of tritium fuel per day. ITER is not particularly telegenic at the moment because it is under construction and will not be commissioned until 2019. This is why we chose to focus on the US National Ignition Facility, which is already up and running.

NIF is pure science fiction; in fact, it was used as a set for *Star Trek: Into Darkness*. It is the world's largest laser system by an order of magnitude. The laser delivers 500,000 gigawatts of power onto a target smaller than a peppercorn in a series of increasingly powerful hammer blows, tuned to arrive with a precision of better than a tenth of a billionth of a second. That is 1000 times the peak energy-generating capacity of the United States. This, as you can imagine, creates a bit of a bang. The peppercorn-sized target contains deuterium-tritium fuel, just like ITER. The laser pulses raise the temperature of the pellet's gold container, and the X-ray radiation produced drives a rapid collapse of the fuel, initiating fusion. The devil is in the detail; the precise timing and duration of the laser pulses, and the shape of the gold container, all contribute to the chances of success and the efficiency of the process. Despite the tremendous engineering difficulty, in September 2013 more energy was released from a deuterium-tritium fuel pellet than the pellet absorbed, although this was only 1 per cent of the total energy input to the lasers. Nevertheless, this demonstrates that so-called inertial fusion works in principle. The inertial fusion power station of tomorrow would use far more efficient laser systems – NIFs are now more than a decade out of date – and the fuel pellet technology being developed by NIF. The technology has been demonstrated to work, at least on a vast, government-funded research scale, and this is how difficult things like space exploration have to begin. Commercial companies will rarely take such enormous risks, and this means that we, the taxpayers, must pay for the creation of this type of knowledge. As with Apollo, we will be repaid, but the investment horizon is beyond that of the average accountant.

It therefore appears that there is no technical reason why such power stations could not be constructed. There is much research to be done,

HUMAN ENERGY
A NASA satellite image shows energy produced by man – white is light, yellow is oil/gas flares and red is agricultural burning.

but the barriers are likely to be budgetary rather than fundamental; the United States spends more on pet grooming than it does on fusion research. There is a serious point behind that cheap shot. I think one of the primary barriers to progress is education. I am a believer in the innate rationality of human beings; given the right education, the right information and the right tuition in how to think about problems, I believe that people will make rational choices. I believe that if I said to someone: 'Here's the deal. You can have limitless clean energy for your lifetime, for your children and grandchildren's lifetimes and beyond, in exchange for grooming your own cat', then most people would reach for a comb. I have to believe that, otherwise this book is a futile gesture.

DREAMERS, PART 2

The second of our stories couldn't be more different. It involves no high technology and very little cash, but it may have a tremendous impact. Securing the future isn't all about money; it's also about action.

The Svalbard Global Seed Vault is modest and beautiful from the outside. In common with all publicly funded construction projects in Norway, the simple door on an Arctic hillside is a work of art, created by Dyveke Sanne. In the summer, it reflects the eternal Sun. In winter, fibreoptic cables shine in the perpetual night. The doorway leads into a converted coal mine, deep in the permafrost. There are three caverns, each maintained at a temperature of -18°C by a cooling system. The temperature was chosen very precisely; it is the temperature at which seeds metabolise slowly, but do not die. At -18°C, the most hardy seeds remain viable for over 20,000 years. Only one of the caverns is in use; the other two are for the future. Inside, there are over 800,000 populations of seeds from almost every country in the world. All the seeds are agricultural crop varieties – the raw material for and the foundation of global food production. Seeds from America and Europe nestle next to those from Asia and Africa. Syrian seeds, rescued from the recent

recent troubles in Aleppo, the home of a local seed bank, sit beside those from North Korea, South Korea, China, Canada, Nigeria, Kenya, and so on around the world. The vault contains virtually the whole history of human agriculture, stretching back to its origins in the Fertile Crescent all those years ago. Each seed population reflects some choices that were made, some environmental challenge or perhaps simply the taste of a farmer or his village. There are varieties manipulated by multinationals, or carefully cultivated and cherished by isolated tribes. The boxes are food for the imagination, time capsules, the stuff of dreams. They are also of fundamental importance.

Why protect agricultural seeds? The answer is that biodiversity is a very good thing. Life on Earth forms a tangled web, a great genetic database distributed across hundreds of thousands of extant species of animals, plants, insects and countless single-celled organisms. The more species there are, the more data there is in the database, and the more chance the whole biosphere has of responding to challenges, be they from disease, natural or human-induced climate change, loss of natural habitat or whatever. This is obvious. If there are genes somewhere in the great database of life that allow wheat to grow with less water, and the climate becomes more arid, then those genes will be valuable to us. If we lose particular genes, then we lose them for good. Today, fewer

'DOOMSDAY' SEED VAULT
A global initiative, the seed vault is located halfway between mainland Norway and the North Pole, deep inside a mountain on a remote island in the Svalbard archipelago. The purpose of the depository is to store duplicates of all seed samples from crop collections around the world.

than 150 species of crop are used in modern agriculture, and 12 of these deliver the majority of the world's non-meat food supply. There is diversity in the form of different varieties, of course; there are estimated to be more than 100,000 varieties of rice. But the overwhelming majority of crop species used throughout human history are no longer cultivated. They are stored, however, in seed vaults, ready for use if needed. The Svalbard Global Seed Vault is a back-up; our insurance policy, ensuring that even if countries lose their seed vaults through natural disasters, war or simple neglect, then irreplaceable parts of the great genetic database of life will not be lost with them.

The Norwegian government owns the seed vault, but the depositors own the seeds. A charitable trust, the Global Crop Diversity Trust, meets most of the operating costs through an endowment fund. Cary Fowler was the executive director of the Trust during the establishment of the seed vault. He was a pleasure to speak to when we filmed in Svalbard – a dreamer, yes, but a dreamer who gets things done.

'Those of us in my field, we live in a world of wounds,' said Fowler. 'We see the injuries, we see the loss of diversity, the extinction, and at a certain point, enough is enough, and you try to figure out what can we do that's not just stop-gap? That really is long term and that puts an end to the problem of crop diversity. Because we know that we are going to need this crop diversity in the future, it's the biological foundation of agriculture. We'll need it as long as we have agriculture.' Which is as long as civilisation exists, I added. Fowler nodded. 'After that, we won't be bothered, will we!'

The Svalbard Global Seed Vault is built, effectively, for eternity, or at least for tens of thousands of years. It is supported by practically all the governments of the world, and is a genuine investment in our future based on sound science and an understanding of the potential challenges and risks that we may face as a single, global civilisation. It's not big, flashy or expensive, but it's important and, perhaps as importantly, somebody actually did it. I find that inspiring.

So where does all this leave us? All I can do is give you my view. I want to be honest. We didn't set out to make a love letter to the human race when we started filming *Human Universe*. We set out to make a cosmology series, documenting our ascent into insignificance. Things changed gradually as we chatted, debated, experienced, photographed and argued our way around the world, and we realised that, for all our irrational, unscientific, superstitious, tribal, nationalistic, myopic ignorance, we are the most meaningful thing the universe has to offer as far as we know, and when all is said and done, that's a significant thing to be. It is surely true that there is no absolute *meaning* or *value* to our existence when set against the limitless stars. We are allowed to exist by the laws of nature and in that sense we have no more value than the stars themselves. And yet there is self-evidently meaning in the universe because my own existence, the existence of those I love, and the existence of the entire human race means something to me. I think this because I have had the remarkable luxury of spending my time in education. I teach, I am taught, I research and I learn. I have been fortunate. I believe powerfully that we who have the power should strive to extend the gift of education to everyone. Education is the most important investment a developed society can make, and the most effective way of nurturing a developing one. The young will one day be the decision makers, the taxpayers, the voters, the explorers, the scientists, the artists and the musicians. They will protect and enhance our way of life, and make our lives worth living. They will learn about our fragility, our outrageously fortunate existence and our indescribable significance as an isolated island of meaning in a sea of infinite stars, and they will make better decisions than my generation because of that knowledge. They will ensure that our universe remains a human one.

THE END

What a piece of work is a Man. So certain, so vulnerable, so ingenious, so small, so bold, so loving, so violent, so full of promise, so unaware of his fragile significance. Someone asked me what they thought was a deep question: What are we made of? Up quarks, down quarks and electrons, I answered. That's what a Man is. Humanity is more than that. Our civilisation is the most complex emergent phenomenon in the known universe. It is the sum of our literature, our music, our technology, our art, our philosophy, our history, our science, our *knowledge.* I have a recording of Mahler's Ninth Symphony conducted by Bruno Walter made on the eve of the Anschluss. It is suffused with threat. Walter and the Vienna Philharmonic knew what was coming. Hope fades with the last vanishing note, which Mahler marked *'ersterbend'* – 'dying' – in the manuscript. It is Mahler's farewell to life, presaging Old Europe's farewell to peace. None of this depth is present in the physical score itself; those black ink dots on white paper can be digitised using a scanner and stored in a few kilobytes on a mobile phone. The fathomless power of the recording emerges from a finite collection of bits because the performance contains the sum total of the fears, dreams, concerns and anxieties of a hundred lives, played out against a backdrop of a million more. The personal history of each of the musicians, the conductor and the composer, and indeed the history of civilisation, hangs upon the supporting framework of the notes, resulting in a work of infinite complexity and power, because each human being *is* possessed of infinite faculties, emergent from a finite number of quarks and electrons. Our existence is a ridiculous affront to common sense, beyond any reasonable expectation of the possible based on the simplicity of the laws of nature, and our civilisation is the combination of seven billion individual affronts. This is what my smiling seems to say: Man certainly does delight me. Our existence is necessarily temporary and our spatial reach finite, and this makes us all the more precious. Mahler's great farewell to life can also be read as a call to value life with all your heart, to use it wisely and to enjoy it while you can.

From Brian

To George Albert Eagle:
It's your future, little boy.

From Andrew

To my soulmate Anna, my beautiful children
Benjamin, Martha and Theo, my wonderful mum
Barbara, my brothers Paul and Howard and all of
the 'small creatures' whom I am lucky enough to
have with me in the vastness.

INDEX

Entries in *italics* indicate images.

Fermilab, Chicago 162
Fertile Crescent 151
Feynman, Richard 40, 67
Fibonacci numbers 209
51 Pegasi 86
fingerprints 81, 192–5, *192–3, 195*
'First Cause' argument 170–1
First Chkalovsk Air Force Pilots
    School, Orenburg 122
Fleming, Williamina 30
flight, birth of 68, 119, 252
food, oxidisation of 111
footprints, early human *130,* 131
fossil fuels 260, 261, 264
fossils *108–9,* 110, 127, *127,* 131,
    139, 140, *141,* 143, 149
Fowler, Cary 269
Frail, Dale 83
Friedman, William F. 69
Friedmann, Alexander 57, 61
fundamental forces/laws of
    nature 177, 179–90, *182*

G

Gagarin, Yuri 24, 122, 123, 124,
    124–5, 125, 156
Gaia space telescope 25
galaxy formation models 183
galaxy rotation speeds 183
Galileo 11, 16, 17, 18, *40,* 40–3,
    42–3, 45, 46, 51, 59, 162, 195
Galilean Satellites 42–3
Gamma Cassiopeia 132
Gamma Cephei 134
Gamma Velorum 132
Gamow, George 56, 59
Ganymede 42, 101
gauge bosons 179, 198
gelada baboon (*Theropithecus
    gelada*) 127, 128–9, *128–9*
General Theory of Relativity 33,
    45, 50, *51,* 52, 53, 54, 55, 56, 58,
    59, 61, 170, 186, *186,* 220, 240
geometry 18, 24, 25, 50, 134, 209
Giant Amoeba (*Pelomyxa
    palustris*) 112, *112*
Gliese 445 79
Gliese 581 84, 86, 87
Gliese 581-C 88, *88*
Global Crop Diversity Trust, The
    269
gluons 179, 182, 183
golden ratio 209
Goldilocks Zone 84, 85, 91, 97, 100

Gould, Stephen J. 57
gravity 179, 214
    dark matter and 214
    Earth's rotation and 132–3,
        134, 142
    eternal inflation and 224
    fundamental forces and 179,
        182
    General Theory of Relativity
        33, 43, 45–6, 50, *51,* 52, 53,
        54, 55, 56, 57, 58, 59, 61, 170,
        186, *186,* 220, 240
    gravitational waves 53
    Main Sequence and 98, 99
    Moon and 24, 132–3, 134, 142
    Newton's Law of Gravity 18,
        24, 25, 26, 33, 45–6, 53, 160,
        161
    orbits of three bodies under,
        unpredictability of 246–7,
        249
    quantum theory of 170, 173,
        186, 225, 226
    space travel and 79, 160
    Standard Model and 186
Gray, Tom 131
Great Debate 35–6, 37, 56
Great Pyramid, Giza 134
Great Silence 117
Greeks, ancient 14, 86, 105, 133,
    152
greenhouse gases 261, 264
Grissom, Gus 132
Guassa Plateau 127–9, *128–9*
Guth, Alan 219

H

habitable zone 70, *70–1,* 84, 94,
    94–5, 100, 101
Haise, Fred 145–6, 147
Hamelin Pool Marine Nature
    Reserve, Western Australia 111
handprints, early cave 232, *232–3*
Haplorhini suborder 127
Harriot, Thomas 198
Harvard College Observatory 30
'Harvard Computers' 30
Harvard University 57
Hat Creek Radio Observatory 70
helium 33, 98, 179, 214, 260
Henry III of France, King 11
Hertzsprung-Russell diagram 97,
    98, 99, 100
Hertzsprung, Ejnar 32, 33, 97, 98,

99, 100
Hewlett Packard 70
hieroglyphs *162–3, 163–4*
Higgs Boson 179, 182, 183, 186,
    222–3, 232
Higgs Field 183, 186, 214, 220
High Andes 94, *94–5*
Hipparchus 133–4
Hipparcos (High Precision
    Parallax Collecting Satellite)
    26, *27*
Hiroshima, nuclear attack on 68,
    117, 234
Holmes, Sherlock 192
Holocene period 151
hominids, evolution of 126–43
Homo erectus 126, 140, 149
Homo floresiensis 149
Homo habilis 126, 140
Homo heidelbergensis 126, 140
Homo sapiens 68, 106, 119, 126,
    127–8, 140, 149, 233
homogeneity and isotropy,
    assumption of 57
Honeysuckle Creek Tracking
    Station, Canberra, Australia
    145
Hooker telescope, Mount Wilson
    Observatory, California *34,* 35,
    36, 37
Hopkins, Mike 157, 166
Horizon: 'The Pleasure of Finding
    Things Out' 40
Horsehead Nebula 30
Hoshide, Akihiko 239
Hoyle, Fred 173, *173*
Hubble Space Telescope 32, 59, 99,
    218, 219
Hubble, Edwin 30, *34,* 35, 36, 37,
    56, 59
Hubble, John 36
Hubei Province, China 127
Human Universe 9, 46, 72, 118, 127,
    157, 239, 251, 260, 269
humans:
    evolution of 35, 106, 110, 111,
        112, 113, 114, 115, 116, 117,
        119, 126, 127–8, 138–43
    future of 230–71
hunter-gatherers 149, 152
Huxley, Thomas 35
hydrogen 68, 69, 70, 74, 79, 81, 94,
    98, 99, 111, 145, 179, 202, 214,
    260, 264
hydrogen bomb 68, 234

LM

LUNAR MODULE (LM)

PICTURE CREDITS

ACKNOWLEDGEMENTS

We first began to discuss the television series that became *Human Universe* in the summer of 2012. It's the fourth major television project we have worked on together and like all of the previous series it has required the talent and dedication of a brilliant team of people. We'd like to thank them all for the endless passion and commitment they have given to the series. We'd especially like to thank Gideon Bradshaw, the Series Producer, for his outstanding leadership. Gideon has worked on many of our television projects over the years including *Horizon* and the *Wonders* series and as always his creativity, vision and passion have been ever present during the production of *Human Universe*. The team also consisted of a world-class group of directors: Stephen Cooter, Nat Sharman, Annabel Gillings and Michael Lachman. The ability to take complex scientific ideas and transform them into beautiful films is a rare talent and we are lucky to have had such expertise on the project. We would also like to thank the hugely talented Paul O'Callaghan, Director of Photography, who has brought such a vibrant beauty to the cinematography across the series, Andy Paddon, 'soundman', for his endless hard work across all of the films, Rob McGregor for his coffee machine and for shooting so many beautiful scenes both above and below the water and Phillip Sheppard for his beautiful score. We'd also like to thank Davina Bristow, Mags Lightbody, Laura Flegg, Alice Jones, Jodie Adams, Karen McCallion and Eloisa Noble for all the ideas and dedication they have brought to the series.

Editing is such an important part of the television-making process and we are hugely grateful to Darren Jonusas for his craft in shaping the series along with the other superb editors Graeme Dawson, Louise Salkow and Gerard Evans. We'd also like to thank Rob Hifle and the team at BDH for the design and visual effects they have brought to every film.

Every production also needs a brilliant team back in the office and *Human Universe* relied again and again on the leadership of Production Manager Alexandra Nicolson, Production Executive Laura Davey and the hard work and dedication of all the production team. Thank you to Louisa Reid, Viola Schwedhelm, Carly Wallis, Alexandra Osborne, and all of those who worked so tirelessly to support this complex production. We'd also like to thank Nik Sopwith and Kate Bartlett who helped shape and nurture the ideas that would form the foundation of the series during its early development.

There are of course so many others who helped make the series and we are grateful to them all, but we would like to thank Peter Leonard, Jenny Scott, Professor Nik Lane, Professor Jeff Forshaw, Professor Frank Drake, Martin West, Julius Brighton, Helene Ganichaud and Vicky Edgar.

Plus a very special thanks to Sue Rider for all of her endless support.

As always, the team at HarperCollins have been outstanding in delivering such a beautiful book against punishing deadlines. We'd particularly like to thank Zoë Bather, whose brilliant design is found in this and all three of the *Wonders* books, Michael Gray, Julia Koppitz, Chris Wright, Anna Mitchelmore and of course Myles Archibald, our patient and wise publisher.

We'd also like to thank the University of Manchester for their continued unwavering support and encouragement, in particular Dame Nancy Rothwell, President and Vice Chancellor, who allows her academics the freedom to be academics.

Cognitio, sapientia, humanitas.

William Collins
An imprint of HarperCollins*Publishers*
77–85 Fulham Palace Road
London W6 8JB

WilliamCollinsBooks.com

First published by William Collins in 2014

Publishing Director: Myles Archibald
Senior Editor: Julia Koppitz
Copy-editor: Helena Caldon
Indexer: Ben Murphy

Cover Design: Julian Humphries
Interior Design and Art Direction: Zoë Bather
Layout and corrections: Michael Gray at Essential Works Ltd
Collage illustrations: Darrel Rees
Production: Chris Wright

Colour reproduction by FMG

Printed and bound in Spain by Graficas Estella